T5-CVM-158

The Britannia Bridge

This is Number 10 in a series of monographs in the history of technology and culture published jointly by the Society for the History of Technology and The MIT Press.

Previous publications in the series include:

History of the Lathe to 1850, Robert S. Woodbury

English Land Measuring to 1800: Instruments and Practices, A. W. Richeson

The Development of Technical Education in France 1500-1850, Frederick B. Artz

Sources for the History of the Science of Steel 1532-1786, Cyril Stanley Smith

Bibliography of the History of Technology, Eugene S. Ferguson

The Hollingworth Letters: Technical Change in the Textile Industry, 1826-1837, Thomas W. Leavitt, editor

Science in France in the Revolutionary Era, Described by Thamas Bugge, Danish Astronomer Royal and Member of the International Commission on the Metric System (1798-1799), Maurice P. Crosland, editor

Moving the Obelisks, Bern Dibner

Officina Ferraria. A Polish poem of 1612 describing the noble craft of ironwork, Walenty Rozdaiensky. Edited by Waclaw Rozanski and Cyril Stanley Smith

The Britannia Bridge: The Generation and Diffusion of Technological Knowledge

Nathan Rosenberg
and
Walter G. Vincenti

The MIT Press
Cambridge, Massachusetts, and London, England

This book was set in IBM Bodoni by Techdata Associates and printed and bound by Halliday Litho Corp. in the United States of America.

Library of Congress Cataloging in Publication Data

Rosenberg, Nathan, 1927-
The Britannia Bridge.

(Monograph series–Society for the History of Technology; no. 10)
Includes bibliographical references and index.
1. Britannia Bridge. I. Vincenti, Walter Guido, 1917- joint author.
II. Title. III. Series: Society for the History of Technology. Monograph series; no. 10.
TG64.B86R67 624.4'09429'21 78-4831
ISBN 0-262-18087-1

Contents

Preface

This monograph is a collaborative work in more than one sense. It is, to begin with, the joint product of two authors from entirely different disciplines, economics and engineering. It is also, however, the outcome of a larger collaboration at Stanford University. For the past few years both authors have been members of Stanford's interdisciplinary Program in Values, Technology, and Society. This program has served as a focal point for faculty and students from different academic disciplines who believe, as we do, that a better understanding of technological phenomena is a matter of the greatest importance, and that such an improved understanding can be achieved only by fusing together the insights of several disciplines. It is, we have found, a difficult but rewarding experience. It is difficult because the extent of academic specialization is now so great that we not only fail to speak one another's language, we often fail even to understand why certain questions should be posed or found interesting. It is also rewarding, however, because the mental partitioning that disciplinary specialization imposes upon us is, although immensely useful and productive, highly artificial. It is therefore, in some respects, a liberating experience to move beyond such boundaries and to share the problems as well as the perspectives of another discipline. We heartily recommend the experience.

In writing this monograph, we have incurred many debts. We wish particularly to thank the following people for their valuable comments and criticisms: D. S. L. Cardwell, Professor of the History of Science and Technology, University of Manchester; Paul David, Professor of Economics, Stanford University; James Gere, Professor of Civil Engineering, Stanford University; Nicholas Hoff, Emeritus Professor of Aeronautics and Astronautics, Stanford University; Stephen Kline, Professor of Mechanical Engineering, Stanford University; and Cutler Shepard, Emeritus Professor of Materials Science and Engineering, Stanford University. We are especially grateful to Virginia Mann for her unflagging diligence and care in typing the various copies of the manuscript. We also want to record our thanks to Stanford University for financial support and to Karyl Tonge, librarian in charge of the Central Map Collection at Stanford, for cooperation in making available the railway map included in the pocket at the back of the book.

Nathan Rosenberg
Walter G. Vincenti

The Britannia Bridge

1

Introduction

Industrialization is a highly complex social and economic phenomenon, necessarily involving drastic changes in society's capacity to transform the raw elements of its physical environment into useful goods, structures, and services. Historically industrialization has meant not only a growth in the output of a fixed, qualitatively unchanged collection of goods; perhaps more important, it has involved also the capacity to produce new goods, to produce old goods by new techniques, to put old resources to new uses, and to exploit on a large scale resources that were once unused or exploited on a much smaller scale. An element essential to this process is the acquisition of new knowledge concerning technological possibilities. Although the assertion that an important learning process underlies industrial development will readily command general assent, the sources, form, and content of that learning process still remain shrouded in obscurity.

The purpose of this book is to shed some light on the historical learning process and thus to illuminate one critical dimension of the dynamics of industrial growth. Our central focus is upon a specific historical event, the building of a bridge, but this single event is a paradigm for a much larger class of events that collectively make up the historical process of industrialization.

The essential nature of technological change has been obscured by an excessive tendency to treat it in vague, general, or purely abstract terms. It may therefore prove a rewarding intellectual experience to look at technological change in particular rather than in general, as the outcome of some specific set of concrete human actions and learning experiences rather than as some vague machina ex deo from which our material blessings have been made to flow. Our assumption is that technological change can most fruitfully be examined as a problem-solving activity. As such, it needs to be approached by asking how specific problems are formulated and how they come to command our attention. The marketplace may register a growing scarcity of an important input, as was the case when the rising price of timber led to attempts to substitute coal for timber in industrial uses as far back as the reign of Elizabeth I. The increasing reliance on coal (as well as other mining industries) led to increasingly serious drainage problems in eighteenth-century Britain. The steam engine, originally used solely for pumping, was the product of a long search to find an instrument to deal effectively with the drainage problem. A flooded mine was a forceful signal of the need for a new or improved technology.[1]

Although these examples involve limitations upon supply that in some sense have been imposed by nature and are reflected in the marketplace, such limitations may be social as well as natural in their origin. Indeed the historical episode that we will describe was one where the problem originated in an act of government; that is, in a constraint that the British government introduced into the construction of a large-span railway bridge. It is likely that this particular source of problem formulation will become even more common in the future, along with a growing social concern over the quality of the environment.

Just as it is important to examine the demand for problem-solving skills, so it is equally important to examine the factors influencing the supply of such skills. Inventions, that is, may be thought of as the product of past attempts to solve particular problems. Their appearance on the historical stage therefore requires that we focus attention upon the skills and knowledge necessary to solve these problems. We need to focus also upon the social sources of these skills and knowledge, the manner in which they have been organized, and the mechanisms by which they have been brought to bear upon the specific problems that society has attempted to solve. However, we are interested not only in the sources of inventive activity, but in their consequences as well. We will examine in detail something that has become a characteristic phenomenon of industrial societies: technological innovations developed in dealing with very specific, narrowly defined problems are found to provide solutions to problems that exist simultaneously in many other sectors of the economy. As a result, such innovations turn out to be far wider in their significance and eventual applications than was anticipated by technical specialists when they addressed themselves to the initial problem.[2]

No society, of course, has ever lacked problems. The intriguing question is the enormous variation, over time and space, in the capacity to deal successfully with them. And, of course, what is particularly striking about that portion of Western history that we have come to call the Industrial Revolution was the remarkable improvement in the capacity to solve the cluster of problems dealing with the production and use of power and the provision and exploitation of cheap metals. By the middle of the nineteenth century British industry was already solidly committed, in Clapham's vivid phrase, to "the wheels of iron and the shriek of escaping steam."[3]

Historians have devoted much attention to the innovations that were connected with the development of new power sources and the application of mineral fuels to the production of metals—as the Clapham allusion emphasizes so well. Curiously little attention, however, has been devoted to closely linked and complementary innovations that served to make possible the application of increasingly cheap and abundant metals to a whole range of new and unaccustomed uses. The episode with which we are concerned was one of the most significant steps in this development, a development that was central to nineteenth-century industrial growth. The speed with which new knowledge in the shaping and application of iron became necessary may be conveyed by the figures for the growth in pig-iron output after the fundamental innovations in iron production in the eighteenth century. British production of pig iron, which was not more than 30,000 tons around 1760, was 325,000 tons around 1818, grew to over 1,000,000 tons in the late 1830s, reached 2,000,000 tons in 1847, exceeded 4,000,000 tons in the early 1860s, and doubled again, to over 8,000,000 tons, in the early 1880s.[4] Not only did the total output of pig iron experience a spectacular growth; of even greater importance for some purposes was the fact that innovations in the iron industry were responsible for the increasing relative cheapness of iron in its specifically wrought-iron form. The great malleability and superior strength properties of wrought iron offered an expanding range of industrial possibilities that had not been feasible earlier when the cost of wrought iron was very high.[5]

Even before mid-century the wheels of iron were creating a range of totally new technical problems. Beginning with the opening of the Stockton and Darlington line in 1825 and rapidly accelerating in the 1830s and 1840s, railway construction rose to a peak in 1847 when no fewer than 257,000 men were employed on the construction of 6,455 miles of railway. By the end of 1850, there were 6,621 miles of railway track in operation in the United Kingdom.[6] The laying down of such vast mileage over all sorts of terrain created construction problems of unprecedented magnitude. Among the most complex of the structures required to accommodate the needs of this system were bridges. Although bridges are surely one of the oldest kinds of structures made by man, nothing in their earlier training or experience had prepared civil engineers of the second quarter of the nineteenth century for the demands that the railway network was imposing upon them. The railway required the construction of bridges involving spans of previously unimagined dimensions and with a capacity to sustain loads and vibrations far beyond any earlier requirements. According to Smiles,

The rapid extension of railways had given an extraordinary stimulus to the art of bridge-building; the number of such structures erected in Great Britain alone, since 1830, having been above thirty thousand, or far more than all that previously existed in the country. Instead of the erection of a single large bridge constituting, as formerly, an epoch in engineering, hundreds of extensive bridges of novel design were simultaneously constructed. The necessity which existed for carrying rigid roads, capable of bearing heavy railway trains at high speeds, over extensive gaps free of support, rendered it apparent that the methods which had up to that time been employed for bridging space were altogether insufficient.[7]

The story of the circumstances leading to Robert Stephenson's unique design for the Britannia Bridge across the Menai Straits in northern Wales[8] (and its smaller counterpart, the Conway Bridge),[9] as well as the details and the occasionally harrowing moments of their construction, have been ably presented elsewhere.[10] Many of the details do not, therefore, require extensive repetition here; for our purpose it will be sufficient to use them in a highly selective way. When Parliament passed the authorization bill for the Chester and Holyhead Railway in July 1844, which involved the selection of Holyhead in Anglesey as the port for the London-to-Dublin train connection, the formidable challenge of providing a rail route across the Menai Straits had to be confronted. How Stephenson, who was engineer-in-chief of the railway, was to meet this challenge was left open at that time. Since operation of the entire line would be held up by any delay in the completion of the bridge, however, the matter was of considerable urgency. (The locations of the Britannia and Conway bridges are indicated on the map in the pocket at the back of the book, which is a reproduction of a map from 1851 showing the railway system of Great Britain in the year following completion of the Britannia Bridge.)

In 1826, Thomas Telford had carried a road across the straits with a suspension bridge in an extremely bold and daring conception for the time. He had been driven to this expedient because the Admiralty had rejected the use of a cast-iron arch because of its interference, through the limitation of headway, with the navigation of the straits by sailing vessels. But the possibility of a suspension bridge for the use of a railway was excluded on the grounds that such a design could not provide the necessary degree of rigidity.[11] In any case, the earlier failure of the bridge across the Tees River for the Stockton and Darlington Railway effectively precluded serious consideration of the suspension principle on so grandiose a scale for railway purposes.[12] As in Telford's case, the Admiralty opposed Stephenson's proposal to span the straits by the use of cast-iron arches. The inability to use either cast-iron arches or suspension methods apparently eliminated the possibility of crossing the straits by what were then considered to be conventional means.[13] Edwin Clark, Stephenson's resident engineer on the Britannia and Conway bridges, summarized the formidable problems confronting the designers in the following terms:

> The natural difficulties to be overcome in crossing such a gulf were . . . much increased by the requirements of the Act of Parliament, by which the dimensions of the central pier were limited, and the roadway, as at the suspension-bridge, was to be 103 feet above the water, this clear height or windway being insisted on throughout the entire span. Thus the arch was rejected; scaffolding from below was impracticable; and the navigation was, under no circumstances, to be interfered with. These were the apparently insurmountable difficulties which the engineer had to encounter and to overcome without delay. No existing kind of insistent [*sic*] structure appeared capable of such fearful extension; and the development of some new principle became imperative.[14]

It was under the circumstances of these unusual constraints that Stephenson conceived of a tubular bridge, to be constructed out of riveted wrought-iron plates, and large enough to allow trains to pass through its interior. To Stephenson, "it appeared evident that the tubular bridge was the only structure which combined the necessary strength and stability for a railway, with the conditions deemed essential for the protection of the navigation."[15] Such a structure, however, was totally unlike anything that had been previously attempted. Wrought iron had never before been used on so large a scale. The novelty of both the materials and the design was so great that there was no reservoir of reliable knowledge or experience upon which to draw in determining feasibility and, above all, safety. Much requisite knowledge had to be fashioned for the occasion since it did not exist at the outset.

2

Generation of Knowledge

A Parliamentary committee conducted hearings in May 1845 on the location and nature of the bridge across the Menai Straits, which had been left undecided in the law authorizing the railway. The resulting bill, which received the royal assent the following June 30, committed Stephenson to constructing a tubular bridge at the location of the Britannia Rock, a large rock in the center of the straits (hence the name of the bridge). Stephenson envisioned that construction of the tubes would take place on a platform supported by wrought-iron chains from the bridge towers, much in the manner of a suspension bridge. After construction the chains might either be removed along with the platform or attached to the tubes as a precautionary support. That decision would be made later. Whatever the decision, the tubes would be designed to be sufficiently strong without the chains.

The unprecedented scale and novelty of the bridge can hardly be overemphasized.[16] The structure (figures 1 and 2) would consist of two parallel lines of tubes, one for trains in each direction. Each line was made up of four tubes, supported by three masonry towers (the center one on the Britannia Rock) and two end abutments. The intended open distance of 450 feet between towers (later increased to

1. The Britannia Bridge over the Menai Straits. From C. Tomlinson, ed., *Cyclopaedia of Useful Arts* (London, 1868), 1: 242.

2. The Britannia Bridge showing Telford's suspension bridge one mile northeast. From E. H. Knight, *American Mechanical Dictionary* (New York, 1877), vol. 3, plate 62.

460 feet) was vastly greater than the 31.5 feet of the previously longest wrought-iron span.[17] It was not far different even from the 580 feet of the previous longest span of any kind in Britain, Telford's suspension bridge across the straits. The distance between the side towers and end abutments was 230 feet, making the total length of each line (including the additional portions bearing on the towers and abutments) slightly over 1,500 feet.[18]

The bridge had three novel aspects:

1. Previously built long-span bridges had been of the suspension or arch type, both of which exert horizontal as well as vertical forces at their end supports. The tubes of this bridge would act as simple beams (or girders), which exert only vertical forces on the supports. This aspect, unlike the other two, was a simplification.
2. Although cast-iron beams were being used considerably in various applications, wrought-iron beams had attained only limited utility, mostly in shipbuilding. The Britannia Bridge would be built of this relatively new structural material.
3. Experience with beams of both cast and wrought iron had been limited to cross-sections of I or T shape (with a single vertical element). Moreover this and the other elements of the cross-section tended to be thick and heavy.[19] The Britannia Bridge would be tubular—with a closed cross-section, as in a pipe—and the walls of the tube would be thin.[20]

In addition, the immense size of the tubes called for riveted built-up construction to an extent not before attempted in a bridge. Finally, fabricating the tubes in place at the required height of 100 feet above the straits was a formidable task. Stephenson later wrote, with all these things in mind, "I stood . . . on the verge of a responsibility from which I confess I had nearly shrunk."[21] Nonetheless by envisioning each tube as essentially similar to two huge I-beams placed side by side (thus II) and knowing that methods of analysis for cast-iron I-beams were reasonably well established, he convinced himself that "the principles upon which the idea was founded were nothing more than an extension of those daily in use in the profession of the engineer."[22] Such optimistic oversimplification is not uncommon, and one wonders how far engineering would have progressed without it.

Stephenson's basic task, then, was to design a practical and economical tubular beam of the required great size and of sufficient strength to support its own weight plus that of a railway train.[23] Economy in this case meant mainly that the cross-sectional shape of the tube and the disposition of material should be chosen so as to keep the total amount of material—and hence the weight and cost of the tube—as low as possible. To accomplish this goal required knowledge of the way beams behave under load and the way this behavior is influenced by their form and size. By 1845, the basic theory of how beams resist the bending caused by vertical loads had been accurately established by French engineers and scientists and introduced into England by the lectures and textbooks of Henry Moseley.[24] From this theory formulas existed for calculating the internal stresses in beams of different cross-sectional shape under certain special—and not always applicable—assumptions. It was not at all clear at the time, however, how these calculated stresses should be used in estimating the ultimate strength—that is, the breaking load—of a beam of specified material. Moreover the few experimental studies of this problem that did exist for metal beams were confined mainly to cast iron, a notoriously nonuniform and confusing material. Small wonder that practical engineers put little reliance on theory in the face of such gross uncertainties. Working in this atmosphere, and obviously still uneasy about the tubular form of his beam despite his earlier optimism, Stephenson decided to carry out model tests "prior to finally deciding on the exact dimensions of the tubes or mode of procedure."[25] It is well that he did so.

Stephenson had seen the need for tests even before passage of the parliamentary bill. In April 1845, while he was explaining his embryonic views on the tube to his father, the famous railway engineer George Stephenson, the two men received a visit from Willaim Fairbairn (figure 3). Fairbairn (1789–1874), a prominent shipbuilder and iron fabricator, was also a versatile engineer, experienced in both mechanical and structural testing and design.[26] He expressed confidence in the feasibility of the tubular beam, and the two men decided at once to arrange a series of experiments. In July, immediately after passage of the bill, Fairbairn began the experiments at his shipbuilding plant at Millwall near London. He in turn enlisted the aid of a friend, Eaton Hodgkinson (1789–1861), who had done structural testing of cast-iron beams and columns under Fairbairn's sponsorship at the latter's ironworks in Manchester.[27] Through private study, as well as reading and discussion with John Dalton, Hodgkinson had acquired considerable competence in mathematics, which he put to good use in important contributions to the strength of materials. He was therefore referred to by his contemporaries, including Edwin Clark, as "a mathematician," a practice since copied by most historians. His work, however, was essentially empirical, and today without question he would be called an engineer. Timoshenko in his authoritative *History of the Strength of Materials* refers to him as such. This shift in terminology from the nineteenth to the twentieth century is indicative of a shift in the nature of engineering knowledge from an area in which mathematical sophistication was unusual to one in which it is taken for granted.[28]

From July through October Fairbairn, with Hodgkinson assisting by analyzing the data, ran an extensive set of preliminary experiments. These included twelve tests on tubes of circular section, six on tubes of elliptical section, and sixteen on tubes of rectangular section.[29] Within each family the tubes had a range of values of the geometrical parameters: size of section, thickness of plate, and length of span (up to 31 feet). The experiments were thus an example of the common engineering method of parameter variation.[30] The values for the parameters were chosen partly in a systematic, a priori fashion and partly in a trial-and-error manner as the experiments progressed. The tubes were built up of riveted wrought-iron plates, following boilermaking practice at the outset.[31] They were

(a) Robert Stephenson

(b) William Fairbairn

3. The principal figures in the building of the Britannia Bridge. Stephenson from L. T. C. Rolt, *The Railway Revolution* (New York, 1962); Fairbairn from *Life of Sir William Fairbairn* (London, 1877).

loaded by a weight concentrated at midspan (figure 4), and the weight was increased progressively until the tube failed.

These preliminary experiments were clearly exploratory. At the beginning the engineers could not be certain even of what data were needed. Clark, writing in 1850, expressed what was undoubtedly in the minds of Stephenson and Fairbairn, at least instinctively: "It was necessary to determine what kind of information was required, rather than to pursue any definite course, and to ascertain generally in what manner tubes might be expected to fail, and to what extent their strength might be modified by form."[32]

Fairbairn quickly found that practices for the joining and riveting of plates that were satisfactory for boilermaking were not adequate for structural purposes. He therefore immediately adopted improved methods. More important, he also discovered that in cases where the tube did not fail from extraneous causes, such as poor joint design, it nearly always failed by buckling (permanent wrinkling) of the plates on the upper side of the tube (figure 5). This result was completely unanticipated—in Fairbairn's words, "anomalous to our preconceived notions of the strength of materials, and totally different to anything yet exhibited in any previous research."[33] The experiments were thus of long-range importance as the first encounter with thin-walled structures that fail through elastic instability, that is, through buckling.[34] More important at the moment, however, the buckling problem "threatened temporarily even to frustrate the consummation of Mr. Stephenson's design."[35]

In the course of the experiments, the engineers decided to adopt the rectangular shape and not pursue certain possible advantages of the curved forms. Their reasons, somewhat complicated, had to do with an observed sectional distortion of the circular and elliptical tubes under load prior to buckling.[36] In addition, in the rectangle, as in the familiar I-beam, the top and bottom are clearly and obviously distinguishable from the sides, something that is not true in the circle or ellipse. This makes it easier to distinguish and think about the different structural functions of different portions of the

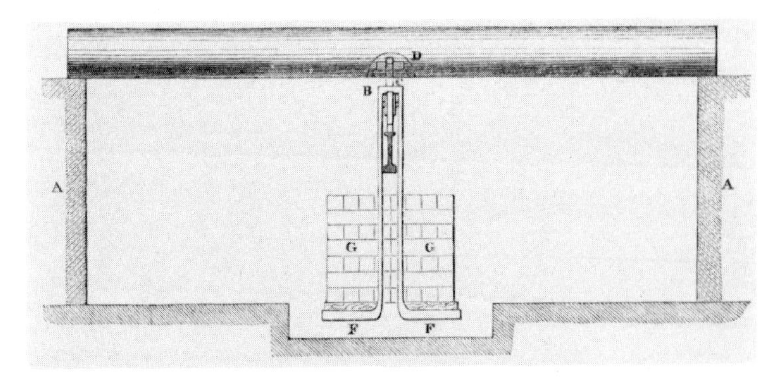

(a) Lengthwise view of circular tube showing load at midspan.

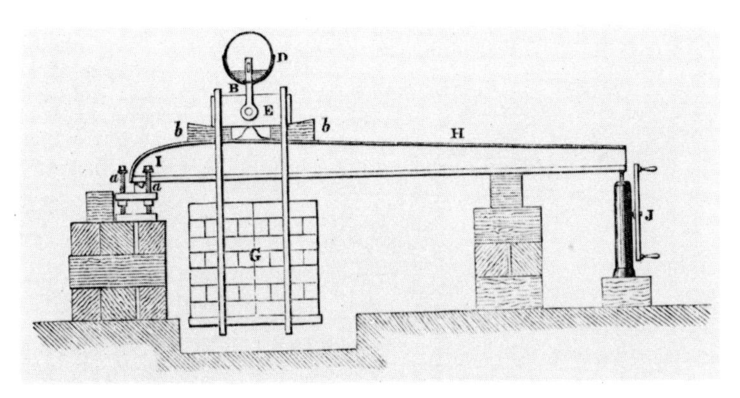

(b) Transverse view of circular tube showing loading device.

4. Apparatus for Fairbairn's preliminary experiments. From William Fairbairn, *An Account of the Construction of the Britannia and Conway Tubular Bridges* (London, 1849), pp. 211–12.

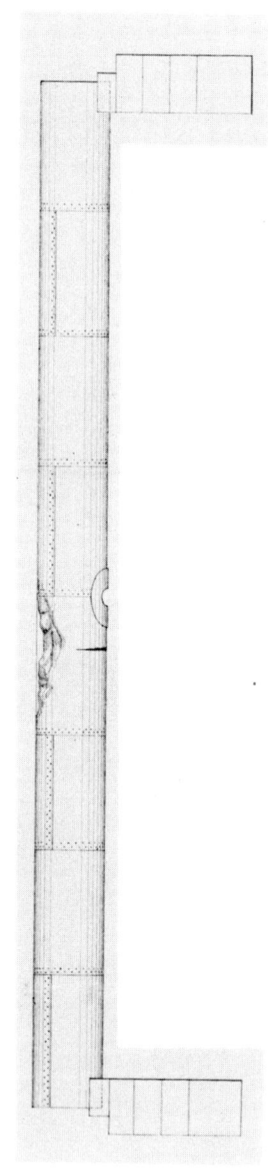

5. Elliptical tube after failure showing both compression buckling at top and tension rupture at bottom; the length of tube between supports is 17 feet. From Fairbairn, *An Account of the Construction of the Britannia and Conway Tubular Bridges* (London, 1849), plate 12.

tube. "The results obtained were consequently more uniform and *intelligible*; and this *simplicity* contributed much to the preference rapidly given to this original form in which the tube had been first conceived [emphasis added]."[37] The intimidating nature of the unforeseen buckling problem no doubt contributed to this decision; with the adoption of the rectangular shape, the crucial difficulty was now confined to a geometrically distinguishable portion of the tube: the top plate. Design engineers, especially in a pioneering effort, often select a particular configuration because it lends itself readily to thought and analysis.[38]

The buckling problem thus quickly became the focus of attention. To make clear what was learned in this regard, we must say something about the way in which a beam resists a vertical load and the extent to which this was understood when the experiments began. Consider the simple case, essentially similar to that of the bridge, of a rectangular tubular beam supported at its ends and subjected to a single load at midspan (figure 6a). This load will deflect or bend the beam elastically in such a way that its top is concave upward and its bottom convex downward, as shown exaggerated in the figure. The theory of beams shows that as a consequence the top portion of the beam—the top flange—is compressed compared with its undeflected length, and the bottom portion—or bottom flange—is stretched. The top flange is thus put in compression and the bottom flange in tension. Now imagine that a center portion of the beam is cut out symmetrically from the rest (figure 6b). It follows from what has been said that the ends of the beam act on this center portion in the same way as would external forces directed as shown by the arrows: the top flange is acted on as if by external compressive forces and the bottom flange as if by external stretching (or tensile) forces. The top flange is thus in a sense equivalent to an isolated bar (or strut) in pure compression, that is, subject simply to a compressive load parallel to its axis (figure 6c). The bottom flange is similarly equivalent to an isolated bar in pure tension (figure 6d).[39] These qualitative theoretical ideas appear to have been fairly well known at the time, and

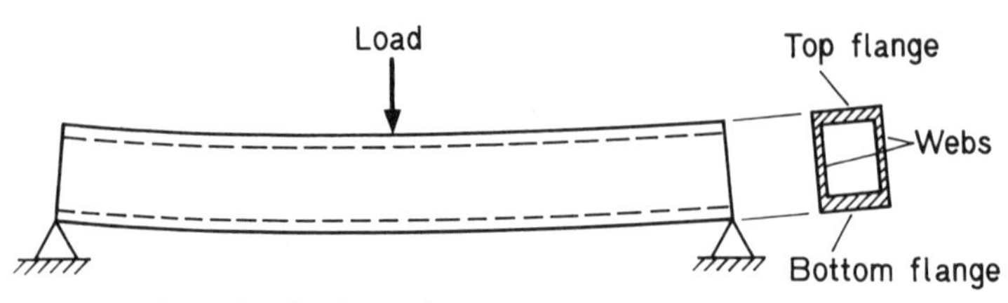

(a) Beam subjected to load at midspan.

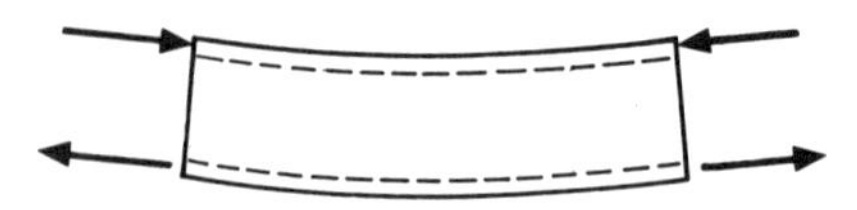

(b) Center portion of beam.

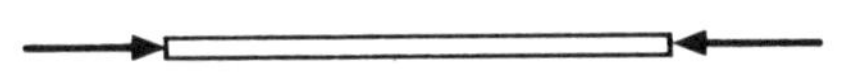

(c) Strut in pure compression.

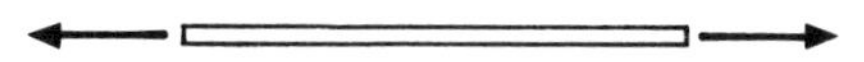

(d) Bar in pure tension.

6. Rectangular tubular beam in bending.

practical methods of calculation based on them and on experiment had been developed for ordinary beams, though they were still hardly precise. Hodgkinson had himself contributed importantly to the state of knowledge in two papers published by the Manchester Literary and Philosophical Society in 1824 and 1831.[40]

A beam subject to bending as just described can fail, in a rough sense, either through failure of the bottom flange as a bar in tension or through failure of the top flange as a strut in compression. Failure of a bar in tension can take place in only one way—by stretching the bar until it breaks, much as might a stretched rubber band. Such failure is a failure of the material. Failure of an isolated strut in compression, however, occurs in different ways depending on the relative length of the strut. If the strut is sufficiently short compared with the dimensions of its cross-section, failure takes place by crushing of the material. This so-called short-strut behavior is thus also a material failure. If the strut is sufficiently long compared with its cross-sectional dimensions, however, failure occurs through catastrophic sidewise bending or flexure—in other words, the strut buckles. Anyone who has pushed hard enough on the ends of a soda straw has witnessed such long-strut failure. Failure by buckling, or elastic instability, is a structural rather than a material failure. The buckling load of a long strut decreases rapidly if the length of the strut is increased; the crushing load of a short strut is a constant independent of the length and is fixed by the inherent strength of the material. The crucial fact to remember is that for struts of the same cross-section, the buckling load of a long strut is always less—usually much less—than the crushing load of a short strut, and hence less than the limits set by the strength of the material. That is to say, whenever a strut fails by buckling it does so at a smaller load than the material itself could bear. Thus, if buckling can be avoided, the load-carrying capacity of a strut can be increased. If it cannot, except perhaps by greatly increasing the cross-sectional area and hence the amount of material, the engineer may be in difficulty. By the time of Fairbairn's experiments, these basic facts concerning the buckling of struts had been well established, mostly on an empirical basis. (The situation vis-à-vis theory was still confused for a number of reasons.) Hodgkinson had again contributed in an essential way to this knowledge in a classic paper of 1840.[41] In this work he provided empirical formulas for the practical design of cast-iron columns like those being used increasingly to support the floor beams of factories and other buildings. For this work he had been elected a Fellow of the Royal Society in 1841 and awarded the Royal Gold Medal. His data and formulas, reanalyzed by others, continued to be used well into the twentieth century.[42]

Considerable knowledge thus existed to help Stephenson, Fairbairn, and Hodgkinson in thinking about the design of a tubular beam. Experience to this time, however, had been with beams of I-shaped section rather than rectangular tubes. An I-beam is in fact structurally similar to a rectangular tube; the top and bottom of the I constitute the top and bottom flanges, and the vertical element provides a single web in place of the two webs of the rectangle (figure 6a). The flanges of I-beams as customarily made, however, were much heavier and thicker compared with the length of the beam than the flanges of Fairbairn's tubes. The compression (top) flange of such beams was thus, in the significant relative sense, like a short strut. Failure of the top flange of previous beams had accordingly been experienced as of the crushing variety (that is, as a failure of the material). The crucial item of knowledge that was thus missing with regard to the bending of tubes—knowledge that the preliminary experiments so startlingly provided—was that the top flange of a thin-walled rectangular tube can behave quite differently. Specifically, the thin plate that constitutes the flange, provided it is long and wide enough that its mid-portion can flex up and down with sufficient freedom, can fail by local buckling. That is, it can fail by elastic instability, much as does a long strut. The thin tops of Fairbairn's rectangular tubes fulfilled these conditions, and they did indeed suffer such structural failure. The beam thus collapsed at a smaller load than it would have if the plate had failed by compressive crushing of the material.[43]

The buckling problem, though at first ominous because it had not been anticipated, was quickly solved in the exploratory experiments. To understand the solution, we must return for a moment to a strut in pure compression. Engineers had known for some time that the buckling load of a long strut of fixed length and fixed amount of material can be increased by making the strut in the form of a tube instead of a solid bar. This strategem works because the tube places the available material farther from the centerline and thus increases the rigidity of the strut against sidewise flexure.[44] In some cases, in fact, the increase in rigidity may be sufficient to cause the strut to fail, not by buckling, but by crushing of the material (as with the short strut). Ideas such as these were undoubtedly behind Fairbairn's testing of three tubes with cellular tops. The third, the final beam in the exploratory experiments, had a rectangular section with the top made of corrugated plates riveted together (figure 7) to form two tubular cells extending longitudinally along the beam. This cellular top was a novel feature at the time. A beam of comparable dimensions but with a flat top had failed, by buckling of the top plate, under a load of 12,188 pounds. The cellular-topped beam gave way by tearing of the sides from the top and bottom, that is, with the buckling of the top suppressed, under a load of 22,469 pounds. Fairbairn considered this outcome "highly satisfactory"[45]—the design principle for the tubular beam had been found. The earlier fears caused by buckling thus quickly proved unwarranted.[46] In individual supplementary reports at the semiyearly meeting of the railway's directors and shareholders in February 1846, Stephenson, Fairbairn, and Hodgkinson told their anxious listeners that the design of the unprecedented tubular bridge could go forward with full expectation of success.[47] (Stephenson's tubular concept and the experiments establishing its feasibility are examined in terms of A. P. Usher's theory of invention in the appendix.)

From the preliminary experiments the engineers had obtained the following results (in more or less increasing order of importance for immediate purposes):

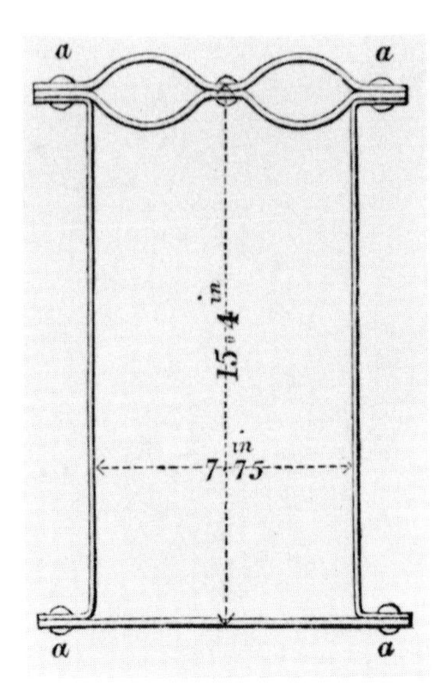

7. Rectangular tube with cellular top flange. From Fairbairn, *An Account of the Construction of the Britannia and Conway Tubular Bridges* (London, 1849), p. 19.

1. Practical information about the best methods of construction.
2. Adoption of the rectangular tube as the preferred form of design.
3. Knowledge of the crucial importance of buckling of the top flange.
4. Demonstration of the effectiveness of cellular design of the top flange as a means to avoid such buckling.

Most important, though least tangible, the engineers had also derived an analytical concept. From point 4, which in turn stemmed from point 2, the top flange could be thought of as a set of familiar compression struts rather than an unfamiliar thin plate. In Stephenson's mind, "The top of the tube thus came to be considered simply as a series of parallel hollow pillars, to resist the compression to which it was subjected by transverse strain."[48] A set of connected cells, of course, differs geometrically from a single strut or column and may therefore differ somewhat in its structural behavior. The engineers fully realized this. Nevertheless by visualizing the upper flange as a set of columns, Stephenson, Fairbairn, and Hodgkinson were now on conceptually familiar ground.[49]

With the basic form for the tube settled, the task became one of deciding on the detailed shape and proportions of the various parts. Ideally these should be such that the top and bottom would fail simultaneously and without buckling of the top. In this condition the material in both parts would be functioning to maximum effectiveness, and the total amount of material and cost of the tube would be minimized. For such refined purposes, the knowledge from the preliminary experiments was clearly insufficient. Even before the reports to the company, therefore, additional experiments were projected or begun. These were of two kinds: tests by Fairbairn at Millwall of a relatively large-scale model of the entire cellular-flanged tube as then envisioned and experiments by Hodgkinson at Fairbairn's plant in Manchester of plates and simple tubes in compression and bending. The purpose of Fairbairn's tests was to arrive as directly as possible at the best proportions—the best distribution of material—through model tests. The purpose of Hodgkinson's was to provide basic knowledge and understanding with the expectation that these would be useful in the full-scale design. These purposes are consistent with Fairbairn's concerns as a practical engineer and Hodgkinson's interests as an analytical investigator.

Fairbairn's tests at Millwall had been projected in December 1845 but were not begun until July 1846. The model was one-sixth scale and thus had a span of 75 feet. Because of the prohibitive cost of building more than one model of such size, Fairbairn here proceeded solely by trial and error: after failure in one test, the tube was repaired, strengthened against the observed failure, and tested again. The cross-section of the model had six square cells in the top flange as illustrated in figure 8, which shows the beam at the outset of the tests. Fairbairn had arrived at this design by reasoning from the

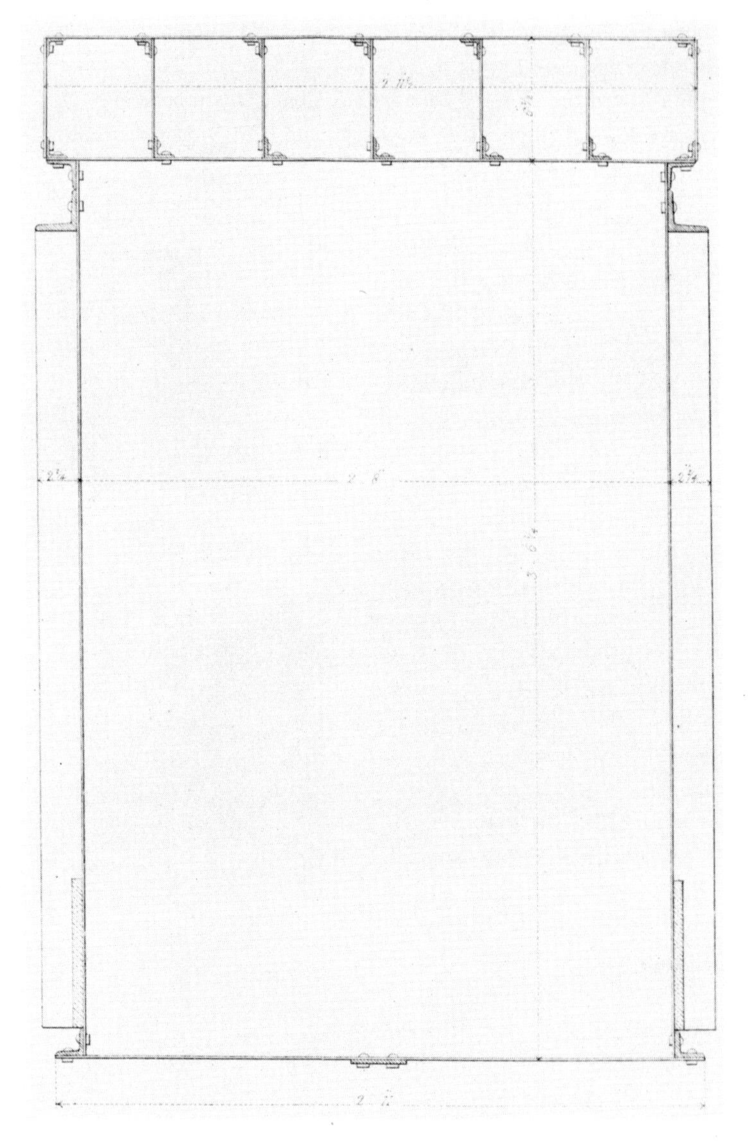

8. Cross-section of one-sixth-scale model. From Fairbairn, *An Account of the Construction of the Britannia and Conway Bridges* (London, 1849), plate 16.

results of the preliminary experiments, particularly those with the cellular-flanged tube of figure 7. The ratio of cross-sectional area of material in the top flange to that in the bottom was 2.6:1.[50] This considerable overstrength of the top was deliberate to ensure that failure would take place in the bottom, which was clearly easier and more economical to repair. The bottom flange was then reinforced in stages as the tests progressed. In the sixth and final test in April 1847, failure took place finally by buckling of the top but under conditions close to the ideal of simultaneous material failure in both flanges. The ratio of cross-sectional areas in the top and bottom had been reduced to 1.2:1. The remaining higher area in the compression flange was a result of the fact that wrought iron fails at a somewhat lesser load per unit area in compression than in tension. Fairbairn had succeeded in increasing the load for failure from 35.5 to 86 tons, an increase of two and a half times, at the expense of only 20 percent more material. In Clark's words, "The magnificent model . . . failed . . . after carrying a greater weight than even a double line of locomotives throughout its entire length. Nothing could be more satisfactory than this result."[51]

Clark's satisfaction is understandable. The model itself had been of a size unprecedented for its purpose. Although the cellular design principle had been established qualitatively in the preliminary experiments, the engineers now had the solid quantitative data essential to the actual design. In fact, the working drawings for the Britannia tubes had of necessity been brought to completion after the next-to-final test of the model. Clark's satisfaction at the final test may therefore have been partly the relief of an engineer who finds that his prior design judgment, on which so much depends, is confirmed. In reporting the success of the experiment to Stephenson, Fairbairn said, "This experiment relieves my mind (as I trust it will do yours) from any doubts as to the proportions, and as to the security of the vast structure on which we are engaged. We have now data on which we can safely depend."[52]

Besides indicating the most advantageous distribution of material, the scale-model tests gave further information about the best arrangements of plates and rivets. Most significant for general knowledge, however, was the notification the tests provided of the importance of the sides or webs of the beam and of the problems that can arise when these elements are thin. To understand this result, we must say more about the mechanics of a thin-webbed beam in bending. Our earlier discussion regarded the top and bottom flanges as operating quasi-independently, that is, as if they were separate members working independently in compression and tension. This view gave no load-resisting function to the web, even though the beam without it would not exist as a single entity.[53] In fact the web is essential to the structural functioning of the beam. In particular, the magnitude of the tension and compression in the flanges varies continuously from one position to another along the beam. The web performs the essential function of balancing the variation in compression along the top flange against the variation in tension along the bottom. As a result the flanges in return impose a so-called shearing load on the web, with the top of the web being pushed toward one end of the beam and the bottom toward the other. If the web is too thin or is unrestrained laterally, such loading can cause it too to buckle, involving phenomena even more complicated than those encountered in the buckling of the top flange. The buckling problem is thus not in fact limited to the top. Although these matters are clear today, hardly anyone understood them at the time. The theory for shearing loads in beams, unlike the well-developed theory for bending loads, had just been worked out by D. J. Jourawski in Russia and was not yet known in Western Europe.[54] Thus no theoretical basis was available for analyzing the catastrophic buckling of the sides that appeared in the second test of the model. Fortunately the difficulty was easily remedied through stiffening the sides laterally by riveting vertical bars to them every two feet across the span. The test let the engineers know, however, that the sides had an essential structural function and would have to be considered carefully in the final design.[55]

Hodgkinson's tests in Manchester were of a different sort. His aim was "to ascertain how far [the buckling of the top flange], which had not been contemplated in the theory, would effect [*sic*] the truth of computations on the strength of the tubes proposed to be used in the bridge."[56] This goal gave him an obviously welcome opportunity to extend his earlier fundamental work on the buckling of columns while at the same time providing results of immediate practical use. The tests included forty plates, twenty-nine rectangular tubes, and thirty-seven circular tubes, all tested in pure compression,[57] as well as thirty experiments on plain (noncellular) rectangular tubes as beams in bending.[58] Since the test objects were simple and relatively inexpensive, Hodgkinson could proceed by a priori parameter variation, that is, by systematically exploring the effects of changes in length, width, plate thickness, and so forth.

The experiments gave a wealth of results. From the compression tests of plates, Hodgkinson obtained the important knowledge that the load at which thin plates buckle increases as the cube of their thickness. A small increase in thickness thus leads to a relatively large increase in buckling load. The compression tests of tubes showed that, other things being equal, circular tubes resist buckling better than do rectangular ones. The bending tests of rectangular tubes verified the prediction of beam theory that, in the absence of buckling, the load-carrying capacity of geometrically similar tubes varies as the square of their length. These tests, which had begun in January 1846, occupied Hodgkinson well into the following year.

As it turned out, Hodgkinson's work was more important for the general knowledge it supplied than for its quantitative use in the design of the tubular bridges. The compression experiments were of particular importance as being, according to Timoshenko, "the first experimental study of buckling of compressed plates and thin-walled tubes."[59] The theoretical study of these problems would not begin until 1891 with the work of G. H. Bryan on the stability of compressed plates.[60] Hodgkinson's empirical work thus provided the basic source of knowledge on the buckling of thin-walled structures for over forty years.

In addition to Fairbairn and Hodgkinson's primary experiments, a variety of secondary tests were made in connection with various aspects of the project.[61] Important among these were measurements of the material properties of cast and wrought iron in tension and compression and of the transverse strength of cast-iron and wrought-iron bars in bending. Knowledge of these properties was still imperfect, especially concerning wrought iron. Also important were tests of the shearing strength of individual rivets, of the failing strength of complete riveted joints, and of the effect on this strength of the contraction that the heated rivets undergo upon cooling.[62] The last matter, in particular, was of unusual interest. The contraction of the rivets squeezes the overlying riveted plates together, and the resulting friction contributes significantly to the strength of the joint. This effect apparently had been unknown previously. Other tests included studies of the resistance of beams to a single transverse impact (or blow) and to continued or repeated impacts. These matters, though not expected to be serious, were of interest in relation to wind gusts and to rapidly moving trains. Finally there were measurements of the strength of stone, brick, slate, and timber. These tests, together with those described earlier, may well constitute the greatest experimental effort in engineering to that time.

The design of the full-scale tubular beams occupied about a year (early 1846 to early 1847). During this period, ideas were examined, changed, and refined–beginning with overall considerations and moving to subsidiary details–until the final design was complete. A preliminary design with considerable detail was finished in July 1846, even before Fairbairn and Hodgkinson's tests were concluded. Since completion of the Britannia Bridge would be the determining factor in the opening of the railway and since its construction would occupy three or four years at least, Stephenson and the directors could afford to wait no longer in letting construction contracts. For economic reasons, therefore, the engineers had to proceed on the basis of incomplete knowledge, a common occurrence in novel and expensive projects.

The design required a series of difficult decisions, interrelated and far from the deductive, linear progression often imagined by people without experience in engineering. The practical requirements of the full-scale tubes, the timing of the design, and the evolving knowledge of the engineers made a simple scaling up of Fairbairn's model inexpedient.[63] Instead the thickness and arrangement of the plates in the top, bottom, and sides were reasoned from Fairbairn and Hodgkinson's results as they became available. Since knowledge was at best incomplete, uncertainty and a great deal of engineering judgment were unavoidable. Disagreement, some of it bitter and acrimonious, was an integral part of the process. Stephenson as the responsible engineer reviewed the major decisions throughout the design, with Fairbairn and Clark carrying out the details. (Clark had been employed by Stephenson as resident engineer in March 1846.) Hodgkinson, at Stephenson's request, made theoretical calculations of the strength and deflection of the Conway tubes in December 1846 and again in March 1847.[64] Clark's book also gives alternative calculations by William Pole for both the Conway and Britannia tubes, but these appear to have been made after the design was completed.[65]

Figure 9 shows the final design for the Britannia tubes; the Conway tubes were essentially similar though somewhat smaller in height and span.[66] The importance of practical requirements in the final design is apparent in the cellular construction of the bottom of the tube (figure 9a). If the bottom had been made in one layer as in Fairbairn's model, this layer would have had to be three inches thick to provide the required cross-sectional area. Since single plates of this thickness could not then be produced, such a layer would have had to be built up by riveting several thinner plates together. Because of uncertainty about the action of the long rivets that would be required, this procedure was not practical, and the bottom as well as the top was made cellular. An additional reason for the cellular bottom in the Britannia Bridge was that in this bridge (though not in the single-span Conway tubes) the four spans in each line were to be

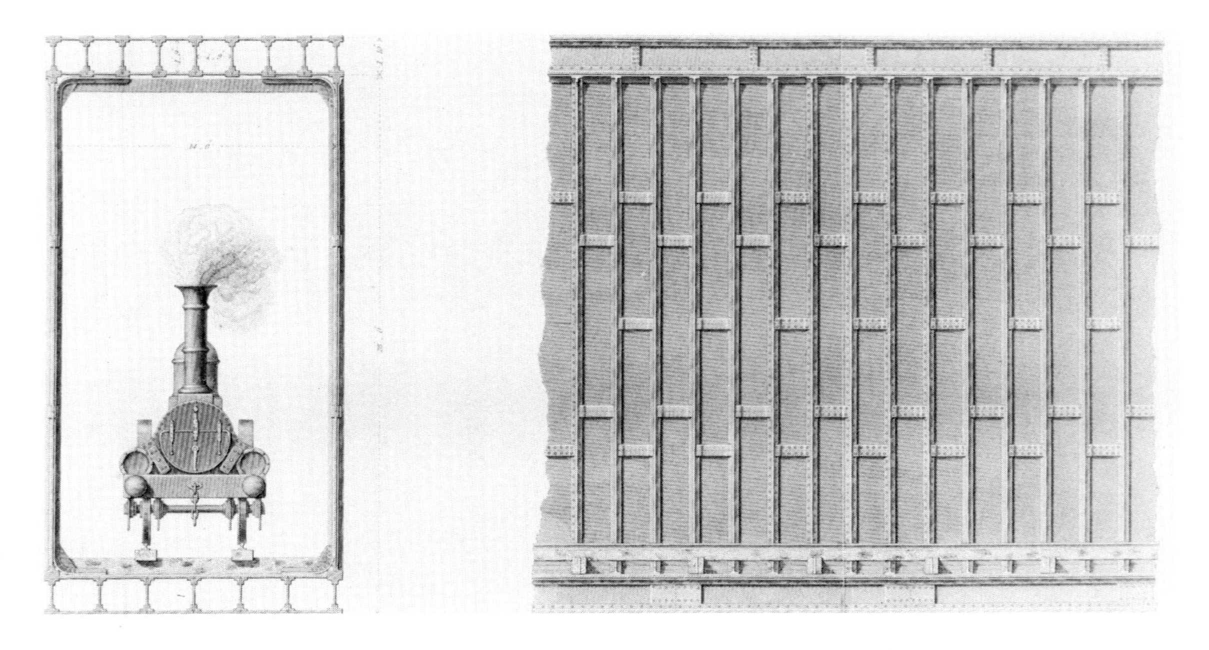

(a) Transverse section through middle of tube.

(b) Longitudinal section through middle of tube.

9. Final design for Britannia Bridge. From Fairbairn, *An Account of the Construction of the Britannia and Conway Bridges* (London, 1849), plate 4.

joined solidly together over the towers to form a continuous beam on multiple supports. This continuity provided additional strength but meant that the beam would bend concave downward near the towers, putting the bottom rather than the top into compression in that vicinity.[67]

How to design the sides of the tube best to avoid buckling was a perplexing problem. Here the engineers had nothing to go on outside the practical knowledge gained from the one buckling failure of the sides in the model tests. Clark, by incorrect reasoning, convinced himself that the vertical stiffening bars that had been used to solve the problem in the model were in fact the theoretically necessary solution.[68] The final design had similar stiffeners made up of two T-irons riveted face to face and covering the vertical joints in the side-plates (figure 9b). Jourawski later made an extensive critique of the bridge design based on his theoretical understanding of shearing loads.[69] He showed that buckling of the sides was caused by internal compressive forces acting diagonally at 45 degrees to the horizontal and that stiffeners inclined in this direction would be considerably more effective even if made with smaller cross-sectional area. Such stiffeners (and the diagonal side plates they would have entailed) would have been inconvenient and expensive, however, and they might therefore have been rejected even if the later knowledge had been available. The actual design, though heavier than it theoretically needed to be, was still perhaps the more economical overall. Most large girders today, in fact, have vertical rather than inclined stiffeners.

Working drawings and lists of materials for the Britannia tubes were completed in November 1846. In these drawings, as in the preliminary design made in July, the top flange was formed of rectangular cells arranged in two horizontal layers, one atop the other, in an arrangement thought necessary to obtain sufficient rigidity. The engineers intended, however, that this design might be modified depending on the evidence from Hodgkinson's tests. In December Hodgkinson proposed substituting circular cells, which he found more effective in his compression experiments with single tubes. Stephenson and Fairbairn rejected this idea because of the practical difficulty of constructing an array of such cells. Hodgkinson's results on plates also showed that with the thickness of plate needed to provide the requisite cross-sectional area, there would be no danger of buckling in the full-scale tube if the cells in the top were replaced by a single plate. This situation, so different from that encountered in the exploratory tests on smaller-sized tubes, was the result of the experimentally observed, thickness-cubed increase in the buckling strength of plates with increasing thickness. The fears generated by the early experiments were thus found to be unwarranted—though it was those very fears that led to the knowledge behind this finding. The single plate needed to do the job, however, would have been four inches thick. Again, as in the bottom, an integral plate of this thickness was not available, and a plate of riveted layers was not practical. With rectangular cells the remaining choice, Hodgkinson's tests indicated that a single layer of square cells, proportioned as indicated by the test results, would be sufficient. One layer would also be considerably easier to build than two. The sequence of decisions here is typical of the trade-offs between simplicity of design, availability of material, and practicality of construction that make engineering an art rather than simply (or "merely") applied science. Stephenson approved the single layer of cells in early March 1847, and final working drawings were made at once. Fairbairn's triumphant final test of the seventy-five-foot model in April thus served to confirm the soundness of the engineers' thinking. It came too late to figure in the detailed design.

Throughout their work Fairbairn and Clark took great care in the location and detailed design of the many riveted joints (figure 9b). They were guided by experience from the various tests of tubes, by the results from the experiments on the strength of rivets and riveted joints, and by their analytical common sense. Again, however, they suffered from their lack of theoretical knowledge of shearing loads. Jourawski in his critique was able to show how the number of rivets could have been considerably reduced if such knowledge had been available.[70]

The uniting of each line of spans of the Britannia Bridge into a continuous beam on multiple supports was a matter of considerable novelty. A single beam somewhat more than 1,500 feet long was new to human experience. Accordingly the engineers gave careful attention to the effect of the continuity on both the strength and deflections of the beam. To guide their thinking, Pole used existing beam theory to calculate the deflection curve that a continuous uniform-section beam placed on five supports would assume as a result of its own weight.[71] Clark made corresponding experiments on a long (33 feet), slender wooden rod, and the results of theory and experiment gave reasonable agreement.[72]

The gigantic tube, however, could not be built as an initially continuous beam and then set down on the supports (as was the wooden rod). Instead the continuity had to be obtained by joining the individual spans together after they were in place on the towers and already deflected separately under their weight. The engineers realized, however, that they could turn this requirement into an asset by using it to reduce the strain on the beam. First, they knew, as did engineers generally, that the maximum strain in a continuous beam as a result of its own weight is less than in separate spans; this was the reason to adopt continuity in the first place. They also understood from Pole's calculations that if the beam is made initially continuous and set on the towers, this lesser maximum strain is located over the towers and is greater than the strain at the middle of the spans. (The maximum strain in a separate span occurs at the middle.) Finally they realized that the maximum strain in the continuous beam could be reduced still further if the strains over the tower and at midspan could somehow be made equal.[73] This equalization they contrived to accomplish by joining the spans successively on the towers in an ingenious way that imposed deliberate strains on the tubes. The required strains were calculated from beam theory together with measurements of the deflection of the tubes as the process proceeded.[74] Although the solution of the problem produced no significantly new structural knowledge,[75] it did provide practical experience with continuous beams at a sophisticated level for the time.

One might expect the engineers would use the strengthening effect of continuity to reduce the thickness of the iron plates needed in the bridge. Instead they chose to make the tubes strong enough to bear their load as separate spans; continuity was used only as an extra factor of safety. There appear to have been two reasons for this choice. First, in any novel project with many imponderables, engineers tend to err on the side of safety. In Clark's words, the bridge's "increase in dimensions and unsheltered position rendered such excess advisable."[76] Even more important, with the method of erection ultimately adopted, the tubes had to carry their weight as separate spans while being brought into position on the towers.[77] Stephenson discussed this requirement in detail in an article he wrote a few years later for the *Encyclopedia Britannica*, in which he appears to be defending himself against the criticism leveled by some engineers that the bridge was overstrength and hence unnecessarily expensive.[78] It is not uncommon in structural engineering to have a temporary condition during construction control significant aspects of the design.

The engineers were less logical in following the implications of continuity in the detailed design. As indicated earlier, continuity puts the bottom of the tube in compression near the towers and the top in tension. This is the reverse of the situation at all points on a separate span, the case that dominated the engineers' experiments and thinking. The implications of this local reversal were recognized explicitly in the design of the top cells where the proportioning of the plates and rivets in the vicinity of the towers was modified to withstand the tensile strain.[79] There is no corresponding indication, however, that the bottom cells near the towers were analyzed as elements to resist compression, a curious omission in view of the attention lavished on the compressive behavior of cells in the top. Clark, after discussing the practical reasons for adopting cells in the bottom, simply mentions the local compressive effects of continuity as an additional reason and leaves it at that.[80] In explaining the detailed design of the bottom, he says that "the bottom of the tube may be regarded merely as a chain of plates"[81] (which is permissible only if it is everywhere in tension, as in a separate span) and proceeds to an exposition of the design on that limited basis. Such failure to integrate ideas completely and consistently is not unusual in engineering—as in other endeavors—when so many of the ideas are new and the participants are working under such pressure.

Wind loads and temperature changes have important effects on bridges, and the designers gave them appropriate attention. They came to the conclusion, contrary to the fears of some, that both loading and vibration from even the greatest wind would have negligible effect. Experience after construction of the Britannia Bridge confirmed this expectation.[82] The unprecedented nature and size of the bridge, however, posed a thermal-expansion problem of unexampled magnitude. The engineers calculated that the largest anticipated temperature variation (75°F) would change the length of the complete continuous tube by as much as eight inches. They provided for this movement by fixing the tube at the central tower and allowing it to expand four inches in each direction, the support at the side towers and abutments being supplied by rollers over which the tube could slide.[83] No new knowledge was required for this solution. The surprises concerning thermal movement came after the bridge was built.

The novel structural problems posed by the tubular bridges were thus solved in the remarkably short time of twenty-one months. Stephenson, Fairbairn, Hodgkinson, and Clark had come a long way in terms of knowledge and design judgment from where they had started. Unfortunately, the present account, in concentrating on the nature and growth of knowledge viewed in the light of present-day understanding, misrepresents an essential point: The entire process appears far tidier and more neatly logical than it actually was. The story was in fact filled with the confusions that typically go with incomplete and developing understanding. Different aspects of the problem appeared critical at different times, and the engineers had to shift their focus accordingly. They frequently had to double back when later findings improved their understanding of earlier problems. At every step they were subject to the pressure of time and to the conflict of doing practical design and basic experiment simultaneously. The list of aggravations could easily be lengthened.[84] In addition Hodgkinson and Fairbairn were on unfriendly terms during much of the work,[85] and Fairbairn grew increasingly unhappy with his position in relation to Stephenson. Finally, embittered, he resigned from the project in May 1848. As usual, the creative process was a highly untidy business. Part of the job of engineering is to live with and organize untidiness.

In retrospect, the nature of the structural knowledge generated in the solution of the bridge problem is clear. In a narrow and specific sense, the engineers learned the following:

1. The controlling importance, for thin-walled structures, of previously unknown buckling phenomena.[86]
2. How practically to avoid buckling of a compression flange by the use of cells (the cellular principle).
3. The problems and importance of the web in a thin-walled beam and how to use stiffeners to eliminate web buckling.
4. The detailed behavior of plates and thin-walled tubes in compression and bending.
5. Improved arrangements for plates and rivets and how to assess the strength of riveted joints.

The engineers thus learned a great deal about the useful properties of thin-walled structures and about how to employ wrought iron as a building material. They also amassed a great amount of quantitative data, particularly with regard to wrought iron, of the sort that designers must have to bring their projects to reality.

In a broader and more general sense, the engineers learned something perhaps more important. By struggling with their problem and forming conceptual models, they learned to think synthetically about the design of an important class of wrought-iron structures. This intellectual framework enabled them to combine empirical data, theoretical understanding, and artful surmise—each limited and incomplete—to attain their practical goal. This too constituted engineering knowledge that could be transmitted to others.

The fabrication and erection of the bridges, like their structural design, posed problems whose solution required developments beyond existing technology. While not as fundamental as the advances in structural knowledge, these developments were significant improvements in technological practice.

The fabrication of the tubes was a task of "vast magnitude" for the times.[87] A good measure of the size of the project is the number of rivets in the Britannia Bridge,[88] which Clark gives as 2,190,100.[89] To punch a large proportion of the required rivet holes, Richard Roberts invented and patented a special punching machine. Roberts was a versatile Manchester machine-tool maker and inventor who had earlier developed the first fully automatic spinning mule for the textile industry.[90] In contrast to previous machines that could punch only one hole at a time, Roberts's device could gangpunch successive rows of holes across one of the wrought-iron plates (cf. figure 9b) as the plate passed lengthwise through the machine. The different number of holes required in different rows was ingeniously achieved by actuating the punches on the principle of another machine from the textile industry, the Jacquard loom. With Roberts's machine "the rivet holes of the thousands of wrought-iron plates used in the bridges were punched much more expeditiously and at precisely equal spacing."[91] For driving the rivets Fairbairn provided a version of the steam-driven riveting machine he had invented in the 1830s for boiler manufacture. His machine was used wherever practicable on the Conway tubes. On the Britannia tubes, riveting machines, though installed, were not used, and older methods of hand riveting were employed. Fairbairn attributes this decision to the contractor's objections that the machines were "both expensive and inconvenient."[92] Clark says that it was due to influence of the workingmen, "who look with jealousy on any machinery which abridges their labour."[93] Even the hand riveting, however, was greatly improved by the introduction of 7-pound hammers in place of the 2.5- to 4-pound hammers that had been standard in boiler work.[94] For heating the unprecedented number of rivets prior to driving, a novel furnace was developed; forty-eight were built and stationed at different parts of the works.[95] By these means and others, riveting technology was transferred from boilermaking and shipbuilding into civil engineering, where it had previously been little used, and was significantly improved in the process.[96]

To erect the Britannia Bridge, a decision remained to be made whether the wrought-iron chains intended to suspend the construction platform should afterward be removed or used to strengthen the bridge. By the middle of 1846 it had become apparent that the cost of the construction platform would be undesirably high; that the tubes could probably be designed, as Fairbairn had been insisting on the basis of his tests, to be safe without the chains; and that an alternative means of erecting the bridge was feasible. For these reasons Stephenson decided in mid-July to dispense with the suspended construction platform and the chains.[97] The wrought-iron tubes for the central spans would now be fabricated on stagings on the beach (figure 10), floated to the bridge site on pontoons, and hoisted to their final place on the support towers by means of hydraulic presses located near the tops of the towers.[98] This procedure meant lifting for each tube the unprecedented weight of 1,900 tons through a height of 100 feet.[99] Two presses that had been employed to lift the lighter and lower Conway Bridge were used together at one end of the tube, and a single gigantic press was built for the other. These lifted the tubes slowly in six-foot increments, with first timber and then masonry underpinning being built up beneath the tubes during each lift. Stephenson had introduced this precaution as a result of a near failure in lifting the second tube of the Conway Bridge. It is fortunate that he did so, for on August 17, 1849, when the first tube had been raised about twenty-four feet, the cast-iron cylinder of the large

10. Britannia tube being fabricated on stagings. From S. Smiles, *The Life of George Stephenson and of His Son Robert Stephenson*, rev. ed. (New York, 1868), p. 450.

press burst violently. The tube, though damaged by the resulting fall, was saved from being destroyed by the timber packing close beneath.[100] The engineers attributed the failure of the press to the fact that the cylinder had a nearly flat bottom, like a shallow dish, which caused a weakness in the cast iron. Stephenson accordingly designed a second cyclinder with an ellipsoidal bottom, and this proved satisfactory.

During construction of the bridges the engineers made careful measurements of the deflections of the tubes under applied loads. Given the tentative practical state of the theory of beams at the time, these measurements also provided a contribution to knowledge. In the view of Todhunter and Pearson, in their remarks in 1886 regarding Hodgkinson and Pole's theoretical calculations for the bridges, "These are probably the most important problems to which the . . . theory of beams was ever, or ever will be, applied. For the Conway Tube the correspondence between the calculated and actual [measured] deflections may be described as the best proof ever given of the close approximation of that theory to fact when it is applied to beams under transverse load."[101] As a result, engineers could use the theory with increased confidence.

In addition to their concern for the deflections caused by loading, the engineers had anticipated on physical grounds that heating and expansion of the side of the tube toward the sun would cause a deflection convex in that direction. They expected, however, that this thermal deflection would be relatively slight. They soon learned differently during the loading tests. The flavor of the situation is given well by Edwin Clark in his testimony to the royal commission investigating railway structures:

The first time we found out about the temperature having such an effect was this; we had put into the tube 200 tons on the previous night, and left it in all night. To our great surprise, when we went to examine the tube, it had gone up nearly half an inch. We looked at the instrument, and at the works, and at the weight, and everything was correct. We could not make it out at all; till at the end we observed, upon looking over our accounts, that invariably when a weight was put in at night, when we came the next morning there was always a little discrepancy; and we used to say that, owing to reading the marks by torchlight, we probably did not read them correctly, and we discarded those marks; but the fact of the matter was this, in the cool of the evening the tube resumes its normal position, but when the sun comes upon it in the morning it again begins to rise; so that those deflections are not so accurate as we could wish. The next tube we test, which I hope will not be long, I intend to take precautions to avoid any such anomalies. I mean to depend upon observations made during the night.[102]

Later tests devised specifically to measure thermal motion showed that the mid-point of the tubes could move as much as 2.5 inches, both horizontally and vertically.[103] This is about one-fourth the size of the deflection caused by the weight of the tube. These unforeseen thermal movements fortunately caused no problem, but they put engineers on notice that they needed to be careful about such things in large metal structures.

Finally, the fabrication and erection of the bridges, like the structural design, depended on new skills as well as new knowledge. Clark at the time put it this way: "Only twenty years ago the Britannia Bridge would have been designed in vain. Not only was there then no machinery and tools for the manufacture and working of such heavy plates, but that intelligent and valuable class of men who carry out such operations . . . was then not called into existence,—they are the peculiar offspring of railway enterprise, and among the most valuable fruits of its harvest."[104] All things considered, it is understandable that the distinguished English economist, William Stanley Jevons, some years later referred to the Britannia Bridge as "our truest national monument."[105]

3
Diffusion of Knowledge

In their efforts to deal with the great novelty of design and material in the construction of the Britannia Bridge, the engineers associated with the project inadvertently generated a mass of engineering knowledge of wide-ranging applicability. The experiments, employing large-scale models, were aimed at determining the strength of tubular bridges, but the results were far more general. They constituted a significant deepening of knowledge of the strength of engineering structures generally. Although this knowledge did not contribute directly to the theory of the strength of materials, it was widely circulated, quoted, and used in Great Britain and on the Continent.[106] Since all sorts of structures in addition to tubular bridges were being constructed out of wrought iron for the first time, the knowledge acquired in bridge building had immediate applications elsewhere and was applied wherever built-up beams and columns were employed for construction purposes and wherever the strength of such structural elements and their riveted joints was an important consideration. Indeed Hodgkinson's researches alone constituted, in Timoshenko's words, quoted earlier, "the first experimental study of buckling of compressed plates and of thin-walled tubes."[107]

Some sense of the significance of the Britannia and Conway experience can be derived from the striking frequency of reference to the bridges in the technical literature of the 1850s to 1870s (and even later). Their construction was obviously a highly visible enterprise, and the learning that took place had widespread effects. Unfortunately the course of these effects is often difficult to trace in detail since engineers prefer to communicate by word of mouth and by observing each other's work—what Derek de Solla Price has called the papyrophobic nature of engineers as opposed to their papyrocentric scientific cousins.[108] Thus the influence was doubtless more widespread than the written record reveals.

The most direct effect, of course, was in bridge building. Stephenson built four more tubular bridges: the Brotherton Bridge across the River Aire on the York and North Midland Railway, two bridges in lower Egypt (one of them across the Nile), and a bridge across the St. Lawrence at Montreal.[109] The last, the Victoria Bridge, was remarkable for having a single line of twenty-five tubular spans with a total length of 6,100 feet, about twice the length of both lines of the Britannia Bridge.[110] Interestingly enough, on both the Brotherton and Victoria bridges (and presumably on the Egyptian bridges, though the record is not clear) the tubes were not of cellular design. Instead the upper and lower flanges were made of one or more layers of plates, the upper flange being slightly arched and, in the case of the larger Victoria Bridge, stiffened on the outside by inverted-T ribs.[111] The engineers had thus come to realize that the use of cells, at first thought essential, was in fact only one of a number of ways to avoid buckling; the choice depended on the circumstances. Completed only nine years after the Britannia Bridge, the even longer Victoria Bridge shows how quickly in engineering the unprecedented can become the accepted. The tubular-type bridge was too heavy and expensive to be successful in the long run, however. For long spans it was soon displaced by the lighter and more satisfactory railway bridges of the truss type, being developed in the 1840s mainly in the United States and Russia.[112]

For shorter spans, girder bridges of riveted-plate construction were highly effective, and here the Britannia Bridge experience had lasting use. Such bridges employed not a large tube with the train passing through it but two or more parallel girders of smaller depth with the trains running in the open between the girders. The built-up girders had various cross-sections, embodying two webs (hollow girder) or a single web (as in an I-beam). In either case the top flange could consist of either a simple plate or one or more tubular cells. The important common element was not the shape of the girder but the use of the newly acquired knowledge and data about riveted wrought-iron construction.

Taking their cue from the Britannia experience, engineers continued for a while to design short-span girders mostly with hollow cross-section. Even before the completion of the large tubes, Stephenson in 1846-1847 built a sixty-foot girder bridge of this type to carry a road over the North-Western Railway. This according to Clark "was indeed the first application of hollow wrought iron girders to the construction of bridges."[113] Fairbairn, in 1846 in the midst of his tubular-beam experiments, took out a patent covering several types of hollow girders.[114] He says that following this he "gave designs, and received orders for more than one hundred bridges in the course of a very few years."[115] In 1847 John Hawkshaw also began erecting hollow-girder bridges that "show every sign of being derived from the experiments for the big bridges."[116] At about the same time John Fowler built a similar bridge, the Torskey Bridge, across the River Trent.[117] Thus even before the Britannia Bridge was completed in 1850, the knowledge produced for its design was already having considerable use elsewhere.[118] The resulting hollow beams were the forerunners of the box girders of welded steel or concrete widely used today in various types of bridges.

In the late 1840s and early 1850s, girder designs became simplified, which usually happens when engineers work with a new idea, and there was a trend toward the simple I-shaped plate girder so common today.[119] The intensity with which girder development was pursued is a measure of the demand that railway construction made for bridges in great numbers. In reference to girder bridges, Fairbairn states that up to 1870 he and his company alone had built "nearly one thousand bridges, some of them of large spans varying from 40 to 300 feet."[120]

I. K. Brunel also designed and ran tests on wrought-iron plate girders with cellular flanges, though it is not completely clear what relationship his work had with the Britannia Bridge experience.[121] In addition his large railway bridges at Chepstow (1852) and Saltash (1859), although not tubular bridges as such, incorporated great wrought-iron tubes as compression members. Brunel was a close friend of Stephenson and was an adviser and witness in the floating of the first tubes of both the Conway and Britannia bridges. His later design of a ship hull was consciously patterned after the bridge tubes. Certainly his thinking about thin-walled construction generally must have been influenced as well.[122] Riveted tubular struts twelve feet in diameter also appear as main columns in the steel cantilever bridge by which John Fowler and Benjamin Baker spanned the Firth of Forth (1890).[123] This great bridge thus also depended on knowledge of thin-walled structures, which had its inception in the experiments of Fairbairn and Hodgkinson.

The use of hollow wrought-iron girders was soon extended from bridges to the commercial buildings appearing in increasing numbers. *The Mechanics' Magazine* in 1860 stated in its news columns:

The employment of iron girders upon the tubular principle in the erection of commercial structures, where the greatest space and light are required, has just received a very practical proof of its efficiency by the completion of the magnificent stack of business premises intended for the joint occupation of the Messrs. Courtauld and the Messrs. Wetford in Aldermanbury. The principal iron girder sustains the whole weight of the private portion of the premises of the former, and is capable, according to the *data* of Fairbairn, Tredgold, and others, of bearing tenfold its present burden. By its means the whole of the business part of the establishment is rendered exceedingly spacious and free from impeding archways or pillars, while it admits of the entrance from above of an expansive flood of the pure northern light, so requisite for the examination of silks, crapes, &c. Light is transmitted to the basement story from the same source, without the employment of an open well-hole, and thus none of the ground-floor space is sacrificed. These are bold innovations, readily appreciated by business men.[124]

Through Fairbairn, whose extensive technical writings exhibit a marked influence from the bridge project, the experience from that project had a broad effect in building design. Giedion credits Fairbairn with a major role in "the decisive change in methods of industrial construction that is to be observed shortly before the middle of the century," including the introduction of wrought iron as a building material.[125] According to Carl Condit, Fairbairn's *Application of Cast and Wrought Iron to Building Purposes*, which drew heavily on the bridge experience, "quickly became the gospel in the British construction industry."[126]

The new technical knowledge acquired in constructing the tubular bridges extended far beyond bridges and buildings. One immediate extension was to the construction of cranes. The growth of industry, the increased scale of productive operations, and the improvements in transportation all created a vastly enlarged need for devices to raise and move articles of great weight and bulk. The conventional cranes in use in the 1840s employed straight-line jibs that could be elevated to an angle of 40 or 45 degrees to the ground. Such shapes rendered it impossible to raise very bulky articles—such as a large bale of merchandise or a marine boiler—to great heights. The articles made contact with the jib well before they reached the top, impeding further elevation. To overcome these limitations Fairbairn, utilizing the cellular principle learned on the bridge project, designed a crane with a curved projecting arm, which he patented in November 1850 (figure 11). These cranes "allowed the article to be raised to a greater height, and at the same time offered greater strength and security."[127] Within a few years massive cranes of Fairbairn's design were being built with the capacity to raise weights of sixty tons to the extreme point of the jib.

11. Tubular crane. From William Fairbairn, *Useful Information for Engineers, Second Series*, 2d ed. (London, 1867), p. 151.

In his book *Useful Information for Engineers*, Fairbairn described his cranes as follows: "The cranes were composed of wrought iron plates riveted together, and so arranged as to give the back or convex side an adequate degree of strength to resist tension, and the front or concave side, which in these cranes was of the cellular construction, a corresponding power to resist compression."[128] In their construction, the technique of chain riveting, which Fairbairn says was "first applied in the tubular bridges in Wales," was also employed.[129] But, much more important, the design principles, upon which construction was based, were a direct and self-conscious application of the knowledge acquired in bridge building. Fairbairn explained: "The crane itself is built on precisely the same principle as the tubular bridge, and may indeed be considered as a curved tubular girder inverted, the top side being the front or concave side of the crane, and the bottom side forming the convex or back part of the structure. Hence it may be described as composed of back plates, side plates, and cell plates."[130]

The bridge-building experience influenced the design and construction of machinery in less conspicuous ways. For example, Joseph Whitworth, the most distinguished contemporary designer and manufacturer of machine tools, discarded the prevailing architectural style of framing in favor of the hollow box frame. According to Rolt: "In the Conway and Menai tubular bridges the civil engineers of his day had demonstrated the strength of the box section and Whitworth adopted it extensively in his machine-tool designs, the effect being to exaggerate their uncompromising, rectilinear appearance. Box-section supports replaced the splayed, columnar or curved legs used by other makers."[131]

One of the most important applications of the newly acquired engineering knowledge was to the building of iron ships. The construction of iron ships became permanently established in the 1830s on the Mersey, the Clyde, and the Thames, and Fairbairn had been one of the prime movers and innovators in the industry.[132] The subject of the transfer of technical knowledge acquired in the construction of railway bridges to the design and building of ships is in fact so large that a detailed account could by itself easily assume monographic proportions. The central point is that the shift from wooden to iron construction in shipbuilding sharply transformed the technical basis of the industry. As a result, in the 1830s, 1840s, and 1850s the shipbuilding industry was invaded by people with this technical expertise—engineers and others, like Fairbairn, skilled in the iron trade. Iron shipbuilding was, to a large extent, the domain of mechanical engineers rather than of the established builders of wooden ships.[133]

The critical problem in the structural design of ships and in the determination of the optimal use and distribution of materials turns upon the analysis of the loads and stresses to which the vessel is subjected at sea. In dealing with this problem, as Fairbairn and other sufficiently astute engineers immediately realized, the experimental information produced for the design of the tubular bridges was directly applicable, that is, an iron ship could be regarded for all practical purposes as an oceangoing tubular beam. In Fairbairn's words: "Vessels floating on water and subjected to the swell of a rolling sea—to say nothing of their being stranded or beaten upon the rocks of a lee shore—are governed by the same laws of transverse strain as simple hollow beams like the tubes of the Conway and Britannia Tubular Bridges"[134] (see figure 12). This realization is a powerful and illuminating one, for "provided we assume such a vessel in its best construction, and regard it simply as a huge hollow beam or girder, we shall then be able to apply with approximate truth the simple formulae used in computing the strengths of the Britannia, Conway, and other tubular bridges."[135]

On the basis of these ideas, Fairbairn severely criticized Lloyd's prevailing regulations for the construction of ships and made recommendations for major changes.[136] He proposed relating the amount of material in the hull to the length as well as the tonnage of the ship, making the structure stronger at midship than at the ends of the vessel, and increasing the amount of material in the deck to

(a) Wave at each end; deck in compression, bottom in tension.

(b) Single wave at center; deck in tension, bottom in compression.

12. Loading of ship's hull supported by waves. From William Fairbairn, *Treatise on Iron Ship Building* (London, 1865), p. 9.

make it equal to that in the bottom of the hull. He also called for the introduction of the cellular principle continuously across the bottom of the hull, as well as in the deck in the form of four more-or-less equally spaced tubular cells running fore and aft. These proposals for altering the prevailing design of iron ships flowed, Fairbairn insisted, directly from the experimental data acquired in tubular-bridge construction and not from any "theoretical considerations."[137] Fairbairn repeated his ideas later in his *Treatise on Iron Ship Building* and spoke with approval of their employment by the British navy in H.M.S. *Bellerophon*, which he described as "the first ship built upon what we consider sound principles."[138]

The greatest shipbuilding innovators of the 1850s borrowed directly from the example of the tubular bridges. In designing the "monster ship," the *Great Eastern* (figure 13), at the time by far the largest ship ever constructed, Brunel introduced the cellular principle, as it had been developed on the Britannia Bridge, for the construction of both the deck and bottom of the ship.[139] Indeed in 1853 he

13. The *Great Eastern* laying the Atlantic cable. From W. H. Russell, *The Atlantic Telegraph* (London, 1866), p. 52.

noted the following characterization of the projected ship, the construction of which was soon to begin: "The ship, all iron, double bottom, and sides up to waterline, with ribs longitudinal like the Britannia Tube."[140] As Robb has remarked, "By virtue of the two bulkheads running fore-and-aft over the length of the machinery and bunker spaces the structure was akin to two 'Britannia Tubes' connected together by the double bottom under the engines and boilers."[141] (See figure 14.)

The design of the *Great Eastern* doubtless also incorporated a great deal of the detailed technical information generated in the construction of the tubular bridges. This point is difficult to document, but from the way engineers draw on each other's work, the influence seems inevitable. This is particularly likely with regard to the improvement in riveting. Apropos of Fairbairn's role in this respect,

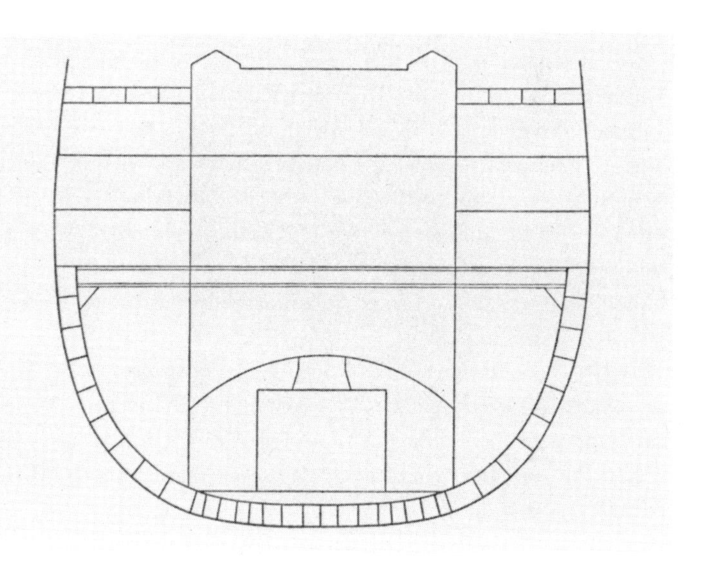

14. Cross-section through the *Great Eastern*. From E. C. Smith, *A Short History of Naval and Marine Engineering* (Cambridge, 1938), p. 110.

Robb observes, "Along with a wealth of general information he had shown that for end-to-end connexion of iron plates a single row of rivets was inadequate, and that at least two rows were necessary. In the *Great Eastern* the butt joints were made with double rows of rivets, whereas all the edge-connexions were made with single rows."[142]

The experiments that gave the information to which Robb alludes had been carried out by Hodgkinson under Fairbairn's direction in 1838, well before the bridge project. The results were not published, however, until 1850 when Fairbairn included them in an important paper to the Royal Society, in which he provided much information essential for the successful construction of wrought-iron ships.[143] The paper makes it clear that in the interim his thinking had been corroborated and influenced by his experience with the tubular bridges. Both Brunel and John Scott Russell, who carried out Brunel's concepts for the hull of the *Great Eastern*, would certainly have been well acquainted with this paper and with the discussion of the experiments on riveting in Clark's and Fairbairn's volumes on the bridges. In his general proposals for improving the design of ships,[144] Fairbairn also recommended the technique of chain riveting used in connection with cranes, which he attributed to his experience with the tubular bridges.[145] Riveting technology, which had been brought into the bridge project from boiler making and shipbuilding, was thus returned, in a much improved state, back to the construction of ships.

The *Great Eastern* not only incorporated a great deal of the knowledge that was generated in the experimental work performed immediately prior to the building of the tubular bridges but was actually built at the same location of many of those experiments—Fairbairn's Millwall facilities, later purchased by Scott Russell.[146] In view of the more than casual resemblance between the Britannia Bridge and the *Great Eastern*, this seems a fitting coincidence.

One final connection between the tubular bridges and the *Great Eastern* must be noted. When the long-awaited day for the launching of this "Great Leviathan" finally arrived, November 3, 1857, its promoters were subjected to the mortifying experience of having the ship, which had been constructed broadside to the river because of the limited width of the Thames, stuck on the ways. The launching was not successfully completed for some months, and only then with the assistance of a massive concentration of hydraulic presses (and at a cost of about £120,000). Among these presses was the same giant press, with a cylinder of no less than twenty-inch inside diameter, that had been used in raising the tubes of the Britannia Bridge above the Menai Straits.[147]

More important in the long run than the specific applications to bridges, buildings, cranes, and ships, however, was the influence of the tubular-bridge knowledge on structural engineering generally. Direct documentation is difficult to come by since design calculations do not often survive and practicing engineers are not in the habit of recording their sources. That the Britannia Bridge experience yielded a great deal of useful structural information, however, is clearly indicated by the many references in contemporary engineering literature. Even before the bridge was finished, Clark, Fairbairn, and Dempsey had published their detailed accounts of its construction. Clark in particular recorded Fairbairn and Hodgkinson's experiments in considerable detail, along with other experiments made in the course of the work. Hodgkinson's experimental findings had already been published as an appendix to the lengthy report by the government commission appointed to examine the application of iron in railway bridges after the failure of a number of such structures.[148] The calculations that Hodgkinson made of the strength and deflection of the tubes were considered so unusual at the time that their details were reproduced in full for other engineers in Clark's book, in the commission's report, and later in Weale's *Theory, Practice, and Architecture of Bridges.*[149] Fairbairn, who was an indefatigable writer and much interested in engineering education, referred to the results often in his numerous volumes on technical subjects, which were highly popular among engineers at the time.[150] Hodgkinson in 1847 was appointed professor of the mechanical principles of engineering at University College, London, and we can assume that his research for the bridge figured in his lectures. These various media, plus numerous talks to engineering societies and other interested groups by Stephenson, Fairbairn, and Hodgkinson, made the knowledge widely available to an eager audience.[151] The knowledge was doubtless widely employed since information on wrought-iron structures, which was in great demand, was otherwise scarce. Data from Clark's book were used, for example, in the *Minutes of the Proceedings of the Institutions of Civil Engineers* for 1850-1851 in connection with a controversy over the structural design of the Crystal Palace for the Great Exhibition of 1851.[152]

Later textbook writers also disseminated the knowledge. Information from the experiments appears in the books of Thomas Tate (1850), Benjamin Baker (1870), and John Anderson (1872).[153] Undoubtedly the most influential citations were those of W. J. M. Rankine, who did as much as anyone to foster the synthesis of theory and practice in British engineering. His *Manual of Applied Mechanics*, first published in 1858, incorporates material from the tests on beams and columns, and his *Manual of Civil Engineering*, which first appeared in 1862, has a thorough treatment of built-up girders and of the great tubular bridges in general.[154] These books went through many editions (twenty-one in the case of *Applied Mechanics*) and were used well into the twentieth century. As late as 1921 Hodgkinson's results on tubes in compression appear in E. H. Salmon's *Columns.*[155] Outside Britain the construction of the tubular bridges attracted great interest, and the results of the experiments were widely used and quoted in the engineering literature. Karl Culmann in particular was impressed by the newly completed bridges and discussed them in his writings, which "greatly influenced the growth of theory of structures and of bridge engineering in Germany."[156] The contributions from the bridge experience were also "frequently referred to [in France] by Saint-Venant in his edition of Navier's *Leçons.*"[157] The results of the bridge experiments thus found their way into the general stream of knowledge of structural engineering.

Numerous references appeared in the more popular technical literature. Tomlinson's *Cyclopaedia of Useful Arts and Manufactures*, for example, devoted thirteen pages to a detailed account of the Britannia and Conway bridges.[158] Later editions of *Ure's Dictionary of Arts, Manufactures, and Mines* contained an article concerned incidentally with the Britannia and Conway bridges but mainly with the hollow-girder bridges that Fairbairn built on the basis of his patent.[159] Stephenson himself wrote about the great tubular structures in his extensive article, "Iron Bridges," in the *Encyclopaedia Britannica*.[160] In the United States articles appeared in a number of places, including *Knight's American Mechanical Dictionary*.[161] Anyone wanting to learn about the technical features of the tubular bridges could thus readily do so.

In addition to producing knowledge directly in the course of their construction, novel engineering projects—successes as well as failures—also stimulate it through ex post facto analysis and criticism. We have already mentioned the critique of the Britannia design by Jourawski in Russia, which furthered the understanding of web buckling caused by shearing loads and of the optimum use of rivets. With this later theoretical knowledge Stephenson and Fairbairn could have built a better bridge, but without it they built a reasonable and sound one—a typical situation in the progress of engineering knowledge. Clapeyron in France also criticized the Britannia Bridge with regard to its properties as a continuous beam on multiple supports. He was mistaken, but his concern was indicative of the growing understanding of continuous beams, in which the Britannia Bridge played a role.[162]

Although the experience with the Britannia and Conway bridges produced a great advance in the knowledge of the strength of engineering structures, it did not contribute significantly to the theory of the subject. Statements are sometimes made to the contrary, but without substantiation.[163] Timoshenko, discussing British experimental work on the strength of materials, of which the work on the tubular bridges was a prominent part, says, "Although this . . . work did not contribute much to the general theory of strength of materials, it was of great use to practical engineers in providing answers to their immediate problems."[164] Engineering knowledge takes various forms, and we must continually remind ourselves that knowledge does not necessarily require or even imply theory.

Most of what we have said has been concerned with specific applications or with the dissemination of specific structural knowledge. In a sense this is secondary. From a broad point of view, the most momentous influence of the tubular-bridge experience, economically as well as technologically, was its role in the shift from cast iron to wrought iron as the material of choice for engineering structures. The history of this important shift has been well traced by Sutherland.[165] He begins:

> In 1840 cast iron was firmly established as the modern structural material and when people talked of "iron" in bridges and buildings they meant cast iron. At this time wrought iron was rare and, for beams virtually unheard of, yet less than 10 years later no engineer would have thought of making a major beam or girder of anything else. By 1850 cast iron had been completely eclipsed. This was not just a time of change from one material to another, but also of unprecedented advances in our understanding of their use.[166]

The change began, of course, well before 1840. Once the introduction of the puddling process by Henry Cort in 1784 had made increased quantities of wrought iron available, it began to be rolled into plates that were riveted together to make boilers and ship hulls. In the early nineteenth century there was also a small but increasing market for wrought iron in chains for suspension bridges, as tension members in roof and bridge trusses made otherwise of timber or cast iron, and as "helpers" in cast-iron beams (rather like steel bars in present-day reinforced concrete). In all these structural applications, wrought iron was thought of as suitable only for resisting pure tension. "The great step forward in the middle 1840s was the establishment of wrought iron by itself as a material to resist bending."[167] This too had its antecedents,[168] but the pivotal technological event was the construction of the great tubular bridges. The experience gained on these bridges showed unequivocally that wrought iron could be fabricated successfully into large riveted structures to resist all manner of loads. To a degree unapproached in any other project, it produced the knowledge of how to use this increasingly available material. It taught engineers how to think synthetically about the design of wrought-iron structures, and it supplied them with a wealth of the quantitative data essential for such design. In short, in solving their unprecedented problem, Stephenson, Fairbairn, and Hodgkinson had "established wrought iron on a more rational basis than any other available material."[169] In addition to its applications in bridges, buildings, cranes, and ships, wrought iron was used increasingly in boilers, steam engines, and wherever else a ductile, high-strength material was required. The experience on the tubular bridges had "proved wrought iron so fully that the technique of its use remained almost unchanged until well into this century, even though the material itself had been replaced by steel."[170]

Nothing we have said concerning the role of the Britannia Bridge in the history of engineering is intended to imply that the historical path that it provided was, in some significant sense, unique or indispensable. The requisite knowledge that it generated for the use of wrought iron in a wide range of constructional purposes would surely have come anyway, though more slowly and along different paths. It is highly likely, however, that the high degree of visibility of this particular bridge-building enterprise resulted in a more rapid diffusion of the knowledge than would have occurred if it had emerged from several smaller-scale undertakings. Although the unusual bridge design had limited commercial applications, the social benefits of the technical knowledge that it generated were widespread. These benefits are, conceptually, measured by how much sooner the useful knowledge was made available than would have been the case had the bridge never been built. We do not presume to offer an estimate of the economic value of these highly diffuse benefits, but a compelling case could readily be made that the benefits attributable to the earlier availability and the more rapid diffusion of the new engineering knowledge were large indeed.

One matter remains to be clarified with regard to cast iron and its use in hydraulic machinery. The use of hydraulic presses to raise the tubes of the Britannia Bridge resulted in the acquisition and diffusion of important knowledge concerning the strength of cast iron when it is used for hydraulic cylinders. The immediate occasion was the accident that occurred when the bottom of the cylinder of the large hydraulic press burst during the raising of the first Britannia tube. The cause of the failure was a subject of intense interest, especially in view of the urgent need for a new cylinder to complete the job. According to Clark, "Some argued that no cylinder of that magnitude could be depended upon, that in so great a thickness of metal the contraction and crystallization of the central portions could not be avoided, and that no cylinders would stand so great a pressure for any great length of time."[171] Stephenson and his associates concluded, however, that the failure was caused by a weakness of the cast iron, which was in turn attributable to the shape of the bottom of the cylinder. In particular they postulated that nonuniform cooling and contraction near the junction of the nearly flat bottom and the sides during production of the very thick casting had resulted in nonuniformity and possible cooling cracks, with resultant weakness in the metal at that point. On the basis of this conclusion, "Mr. Stephenson consequently decided on ordering a new cylinder in every respect similar to that which had failed, excepting that the mass of metal at the bottom was reduced in bulk, and the shape of the bottom was modified."[172] This new cylinder, in which the almost flat bottom was replaced by a deeply rounded ellipsoid, was cast and delivered six weeks later, and the tube was lifted without further incident.

The success of the new design, especially in view of the dramatic and highly visible circumstances of its introduction, caused considerable discussion among engineers at the time.[173] Two decades later John Anderson was still moved to give an alternative explanation, this time in terms of crystal formation rather than possible cooling cracks. In his influential textbook on materials and structures, he wrote:

The hydraulic cylinders were originally made with a flat bottom, like that of a drinking-glass [figure 15a];[174] the cylindrical part of the casting had the crystals radial from the inside, but in the bottom part the crystals were perpendicular to the flat end, and at the points where the two different arrangements of crystals come together, at an angle—namely, at a line drawn from the inner to the outer corners—there were the lines of weakness. Hence it was that, although considerably thicker, the cylinder failed at those points.

The second or substitute cylinder was made with a hemispherical end [figure 15b], and in it the radiating crystals were all arranged in lines more nearly parallel, although of course not truly so. As thus made, it was found to be amply sufficient in strength, even with the same amount of metal in the mass. The foregoing accident was the means of drawing public attention to the subject. Many engineers found it to agree with their former experience, and many had previously been modifying the forms of structures without knowing the natural law.

. . . Already most of our engineers are constructing their hydraulic cylinders on the plan shown [figure 15b], and the Americans are constructing their cast-iron guns in form not unlike a soda-water bottle, which is nearly in strict accordance with this law of crystallisation, the exterior surface being arranged so as to invite the heat outwards in a nearly uniform current in all directions.[175]

In light of present-day knowledge, either Stephenson's or Anderson's explanation could be correct, though Stephenson's is the more likely. (To settle the matter one would need, not the existing documentary evidence, but an opportunity for an expert to examine the failed cylinder.) Whatever the explanation, the practical lesson regarding the shape of the cylinder was an important one for the design of hydraulic machinery. In addition the lifting of the bridge tubes and the great notice it aroused helped encourage the use of hydraulic-power techniques generally. A contributing factor in this was the exhibit of the largest of the presses at the Great Exhibition

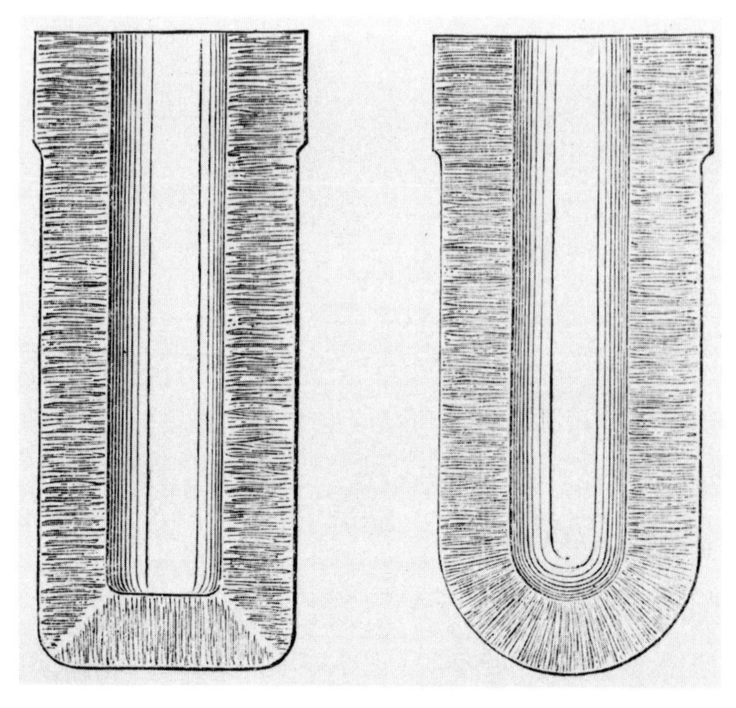

(a) Original cylinder.

(b) Replacement cylinder.

15. Section through cast-iron cylinders of giant hydraulic press. From J. Anderson, *The Strength of Materials and Structures* (New York, 1872), p. 280.

of 1851 (figure 16), where its great size and notoriety made it a focus of public attention.[176]

16. The giant hydraulic press used to raise the Britannia Bridge being exhibited at the Great Exhibition of 1851. From *The Illustrated London News* (Sept. 20, 1851), 19: 380.

4
Conclusion

The problems that emerged in the construction of the Britannia Bridge eventually generated a wealth of technological knowledge, whose benefits extended far beyond the ability to span the Menai Straits for the Chester and Holyhead Railway. Those straits were spanned by an entirely novel design, a tubular bridge. But the tubular bridge itself had only a limited useful life in subsequent years, and in this sense merits not much more than a footnote as far as the history of bridge building is concerned. It was soon displaced for long-span railway bridges by far more satisfactory types constructed along different principles—especially the truss bridge in its various forms. Nevertheless the brief experience with tubular bridges was of far-reaching importance because an extremely fruitful learning experience took place in the process of design and in dealing with the numerous technical problems that such bridges posed. This experience provided knowledge of great value to a wide array of productive activities where wrought iron was employed as a building material. Such knowledge long outlived and vastly exceeded in importance the specific bridge-building circumstances that originally gave rise to it. Indeed it would be difficult to conceive how riveted wrought iron could have become the dominant structural material as early as the 1850s, as it did, without the vital information provided by the bridge-building experience.

These observations contain an important implication that is easy to lose sight of when assessing the historical past from the perspective of the present—and it is, of course, difficult to liberate ourselves from the confining perspectives of the present. It is deceptively easy to evaluate the significance of individual technological improvements in terms of their contribution to that ongoing stream that directly flowed into visible, present-day technologies. If this approach is not successfully resisted (and it usually is not), technological history then becomes the celebration of a series of such success stories. In fact, as we have argued, there was a vast accretion to the stock of technical knowledge and much growth in technical expertise as a result of efforts expended in the development of specific technologies that for a variety of reasons were discarded or bypassed in the later onward rush of industrial development. It is therefore likely to be seriously misleading to write technological history in terms of developments that have survived in present practice. If we are concerned, as we believe it is necessary to be, with the collective learning process underlying industrial societies, then we must recognize that the success and survival of individual technologies do not provide an adequate guide to the sources and the growth of industrial skills. Indeed one of the morals of our story is that the generation of valuable technological knowledge is not confined to episodes that resulted directly in the production of inventions that later proved to be of great value. Much valuable technological progress has taken place on projects that have low visibility now because they did not eventuate in a highly significant invention that we know and use today. Tubular railway bridges, which were quickly superseded by other bridge designs, are an important case in point.[177] During their brief eminence, they were responsible for major improvements in technological knowledge before they faded from public and professional concern. As a result historical accounts that proceed from one presently evident technological triumph to the next are flitting from iceberg tip to iceberg tip, while the larger, frequently anonymous steps in the growth of technological knowledge have taken place, concealed from our present view, beneath the surface.

If this view is correct, our understanding of the historical growth of technological knowledge and skills may be vastly expanded by identifying and studying other specific events that, even though they eventually proved to be commercial failures or experienced only short productive lives, made important contributions to technological knowledge. Indeed the construction of the *Great Eastern* was probably another such event. If we examine technological history by focusing not on the major long-term success stories but upon the historical incidents that resulted in significant accretions to the stock of human knowledge, our perspective changes drastically. This new perspective requires that we probe more deeply into the cumulative growth of knowledge underlying technological improvements and examine more carefully the contributions underlying the growth in the stock of knowledge that made specific inventions possible.

In addition there is often an interdependence between technologies such that the limitations of one technology impose restraints, in some cases only temporary, upon others. Such interdependences play a vital role in the timing and the direction of the inventive process. The history of the Britannia Bridge effectively illustrates this point. The bridge assumed its tubular form because of Admiralty restrictions that were designed to avoid interference with navigation of the straits. As a result a projected arch bridge was firmly rejected. But the present-day replacement for the Britannia Bridge is, in fact, an arch bridge.[178] The Admiralty's concern, which was perfectly legitimate with respect to the navigational technology of the tall-masted sailing ship, became irrelevant when new forms of propulsion were substituted for sail.

At the same time the entire bridge-building episode, shaped as it was by Admiralty restrictions, serves to underline the multiplicity of ways in which forces external to the technological realm exercise an influence in the development of new technology. The tubular bridge, and the vast knowledge resulting from it, emerged when it did *not* just in response to the unfolding of some inner technological imperative but also as a result of the way certain institutional forces at the time constrained the choices open to the engineer. There was nothing unusual in all this. Both the path and the timing of technological innovation always reflect a complex interaction between technological possibilities on the one hand and the impact of larger social forces on the other.

The bridge-building experience also exhibits other interesting ways in which separate technologies interact and mutually influence one another. The relation of iron-ship building to bridge building demonstrates a recurring phenomenon of technological history: the reciprocal, constructive interaction of technological innovations in separate industries. Although the building of the Britannia Bridge generated a great deal of invaluable technological knowledge for shipbuilding, technological knowledge drawn from iron ships had also played a critical role, at several junctures, in the building of the bridge. Indeed, when Stephenson was brooding over the feasibility of a large, wrought-iron tubular bridge, his confidence in the project was decisively confirmed by information drawn from an accident in the launching of an iron steamship, the *Prince of Wales.*[179] Although the hull of the ship was accidentally subjected to immense strains, it suffered only trivial damage. Stephenson stated: "The circumstances here brought to light were so confirmatory of the calculations I had made on the strength of tubular structures, that it greatly relieved my anxiety, and converted my confidence into a certainty that I had not undertaken an impracticable task."[180] More generally, a major factor in the successful construction of the bridge was the huge fund of knowledge of ironmaking that Fairbairn brought to the project, knowledge gained from his years of experience in the construction of iron ships. As one recent authority has stated: "No doubt the availability of the good-quality wrought iron plates and sections recently produced for ship-building contributed to this idea."[181] But

Stephenson himself had publicly confirmed his reliance on shipbuilding experience in proceeding with the tubular bridges. During a public discussion of the feasibility of such a bridge, he was asked: "There is no experience of a bridge being formed of a tube of this kind; is there?" Stephenson replied: "No, there is no experience of it; nor was there of the iron vessel some time ago. There is now one building by Mr. Fairbairn, 220 feet in length, and he says that he will engage that when it is finished that it shall be put down in the stocks at each end, and shall have a thousand tons of machinery in the middle of her, and it will not affect her."[182]

One of the most striking aspects of the building of the Britannia Bridge was that a great deal of the technological knowledge that was involved in its construction was "made to order" for the occasion.[183] Much of it was simply not available before commencement of the project, and it is difficult not to sympathize with Stephenson's admission, quoted earlier, that "I stood . . . on the verge of a responsibility from which I confess I had nearly shrunk." It was the need to span the straits with a railway bridge, and subject also to the constraints imposed by the Admiralty's insistence that the bridge offer no impediment to navigation through the straits, that led to the exploration for and development of techniques that did not exist at the time the bridge was initially proposed.

The manner in which the requisite knowledge was acquired is interesting for another reason. There is a widespread tendency to regard technology as involving essentially the application of knowledge derived from science. We believe that such a model of engineering activity is seriously deficient on a variety of grounds, and we intend to deal with these deficiencies at a future date. For present purposes, however, what is clear is that, in the case of the Britannia Bridge, much of the requisite knowledge did not preexist and had to be produced on the spot. In other words the work of the civil engineer involved not the application of existing theoretical knowledge but the design and development of techniques that provided, by means of experimental investigations, empirical knowledge where theoretical knowledge was not and could not readily be made available. For example, Hodgkinson's compression tests of plates bypassed a lack of theoretical knowledge in order to establish the buckling properties of thin plates, a problem that was first studied theoretically by Bryan only forty years later.

Such working solutions are by no means confined to engineering in the nineteenth century. A good deal of the design work of engineers since then remains essentially similar. Although engineers obviously draw upon scientific knowledge when it is possible and feasible to do so, their activities are fortunately by no means confined to or circumscribed by such possibilities. Indeed it may be appropriate to regard the completion of the Britannia Bridge as signaling the emerging maturity of the British civil-engineering profession. In this case it demonstrated its capacity to undertake a project of vastly new dimensions, involving the use of materials in entirely new applications and requiring knowledge and calculations far beyond what was available. It acquired this information in an impressively systematic way. The profession thus convincingly demonstrated the resourcefulness needed to generate the new knowledge essential to the completion of a totally novel project, knowledge that turned out to have applications far beyond the context within which it was acquired. The profession showed that it could, in effect, produce complex technological innovations on demand. The ability of engineers to do these things seems to constitute a reasonable definition of professional maturity.

An additional point removes the work of practicing engineers even more from the interface between the professional discipline and the realm of scientific research: engineers are typically engaged in design activities in which they are subject to constraints of no particular interest to scientists. They are necessarily intensely concerned with financial considerations, for example, and their work in this respect is more closely related to economics than to the physical sciences. They are also concerned with optimizing activities that are of no special relevance to the scientist: maximizing the performance characteristics of some component, minimizing the amount of material in a structure, redesigning a process to substitute a cheaper material for a more expensive one, and so forth.[184]

A few final comments need to be added concerning William Fairbairn. It is customary to call Fairbairn a transitional figure in the history of British engineering.[185] Upon his death in 1874 *The Engineer* said of him: "He abolished the millwright, and introduced the mechanical engineer." Fairbairn started out in a modest engineering partnership in Manchester in 1817 in a shed rented for twelve shillings a week and employing a lathe of his (and his partner James Lillie's) own construction, the power source for which was "James Murphy, a muscular Irishman." Fairbairn's transitional character is nicely epitomized in two papers that he delivered in 1849: one on steam engines, which appeared in the *Civil Engineer and Architects' Journal,* and one that he delivered to the Institution of Civil Engineers bearing the title "On Water Wheels and Ventilated Buckets."[186] By mid-century his large engineering enterprises had produced what appears at first glance to be a bewildering array of industrial products: millworks, waterwheels, ships, bridges, steam engines and boilers, cranes, locomotives. Underlying this diversity, however, is a central and unifying theme: a high degree of sophistication and expertise in the employment of iron for construction purposes. In an age of increasingly cheap iron, Fairbairn possessed a remarkable intuition for grasping the constructional possibilities of this newly abundant material, even before the knowledge of the material had become formalized and codified. In this sense he was indeed a transitional figure. But what needs also to be perceived in this transition is the way he embodied a newly emerging pattern of specialization. The specialization was not, as in an earlier period, identified with a handicraft or a particular final product but with a material. Wherever wrought iron was to be applied to a new structural or mechanical use, his kinds of skills were invaluable because he understood the material and its limitations, and, more important, he sensed its still-unexploited possibilities.[187] With this kind of expertise he moved readily across traditional industry boundaries, precisely because these separate industries had just come to possess a critical but also still strange new dimension in common. It was this strategic position that enabled him to play such a central role in the transfer of new techniques from one industry to another. Techniques that were successful in the shaping of iron products in one industry were readily transferable to other industries employing the same material—but to do so required the perspective of someone who understood what was common as well as what was disparate to different manufacturing processes or physical structures. It required an eye for essentials and a capacity for abstraction—the perception, for example, that in spite of their totally different functions, a bridge, a crane, and a ship could each be treated for design and construction purposes as if they were simply tubular beams.

There was unfortunately a great deal of acrimony between Robert Stephenson and William Fairbairn and their respective partisans over the proper apportioning of credit between these two eminent worthies. We do not wish to enter the lists here beyond rendering our judgment that the primary credit for the conception and construction of the Britannia Bridge should go to Stephenson and that the primary credit for the transfer and application of the new technological knowledge across industry lines should go to Fairbairn. But it is even more important that we should expand our appreciation for the nature of the larger, social learning process in which each of these people played such a distinguished role.

Appendix: The Tubular Bridges and Usher's Theory of Invention

In his classic *A History of Mechanical Inventions* (first published in 1929), A.P. Usher put forward a theory of the inventive process based on the concepts of gestalt psychology.[188] The episode of the tubular bridges provides an opportunity to test this theory against a specific set of events.

Usher regarded the fundamental question for historians to be that of accounting for the emergence of novelty in history. From this perspective inventions represented one specific form of novelty. Usher was particularly concerned with rejecting the heroic view of invention, which explained the inventive process as the acts of uniquely endowed people, or geniuses. Such an explanation, he felt, was no explanation at all. Although the role of "superior persons" played a role in Usher's account, the principles of gestalt psychology offered a basis for understanding the specific circumstances that gave rise to the emergence of novelty. As Usher expressed it, "The distinction between acts of skill and inventions is suggestively drawn by gestalt psychology. Novelty is to be found in the more complex acts of skill, but it is of a lower order than at the level of invention. As long as action remains within the limits of an act of skill, the insight required is within the capacity of a trained individual and can be performed at will at any time. At the level of invention, however, the act of insight can be achieved only by superior persons under special constellations of circumstance. Such acts of insight frequently emerge in the course of performing acts of skill, though characteristically the act of insight is induced by the conscious perception of an unsatisfactory gap in knowledge or mode of action."[189] For Usher, then, a theory of invention needs to be concerned with carefully specifying the circumstances under which inventions are likely to be achieved.

The central idea of Usher's theory is a four-step schema for analyzing the emergence of novelty, which Ruttan has ably summarized as follows:

(i) *Perception of the problem*, in which an incomplete or unsatisfactory pattern or method of satisfying a want is perceived.
(ii) *Setting the stage*, in which the elements or data necessary for a solution are brought together through some particular configuration of events or thought. Among the elements of the solution is an individual who possesses sufficient *skill* in manipulating the other elements.
(iii) *The act of insight*, in which the essential solution of the problem is found. . . .
(iv) *Critical revision*, in which the newly perceived relations become fully understood and effectively worked into the entire context to which they belong. . . .[190]

Usher was aware that major—or strategic—inventions, such as Watt's low-pressure steam engine, cannot be compressed into this elementary four-stage categorization. They are often a complex synthesis of a number of subsidiary inventions and may have numerous problems and numerous interdependent solutions involved in their development. In an attempt to deal with this complexity, Usher conceived of a major invention as occurring in four steps similar to but less well defined than those of his elementary schema. There may be individual elements of novelty feeding into any, or all, of these four major steps. For example, the mere perception of the major problem may require great insight, or subsidiary insights may be necessary in setting the stage for the major act of insight. (Usher further viewed long-term technical achievement as the cumulative synthesis of successive strategic inventions, but this concept need not concern us here.)

We can now attempt to analyze the tubular-bridge events in terms of Usher's four elementary steps but without trying to fit them into his less well-defined representation of the strategic invention. A possible arrangement follows.

1. *Perception of the basic problem.* As a result of the Admiralty's rejection of an arch bridge and the well-known lack of rigidity of previous suspension bridges, Stephenson perceived that conventional long-span bridges could not provide the wanted railway bridge across the Menai Straits.
2. *First setting of stage.* Stephenson considered various kinds of novel bridges, including a suspension bridge with an enormously stiffened deck in the form of a rectangular wrought-iron tube.[191] The skilled individual here is Stephenson alone.
3. *Fundamental act of insight.* Stephenson realized that the tubular deck of the stiffened suspension bridge can be regarded as a simple beam and the suspension chains as auxiliaries, which might be eliminated entirely. He also realized that for analytical purposes the tubular beam can be looked upon as two I-beams side by side.[192]

At this point we might consider that the essential element of novelty had emerged and subsume the subsequent experiments and ideas all under critical revision. Usher's elementary schema would then be complete and sufficient by itself. Such a view, however, would be asking the notion of critical revision to carry a great deal of freight, probably more than it should. We might therefore adopt the view of F. M. Scherer in his discussion of the Watt-Boulton steam engine and regard the experiments and associated thinking as a development process complementary to but distinct from the process of invention.[193] In the present case, however, it seems more illuminating to continue with Usher's elementary steps:

4. *Perception of first subproblem.* Stephenson quickly perceived that, notwithstanding the similarity of the rectangular tube to a pair of I-beams, the contemporary state of engineering knowledge of beams was insufficient for the wanted design.
5. *Second setting of stage.* The exploratory experiments were initiated at Millwall. The skilled individuals were Fairbairn and Hodgkinson, as well as Stephenson.
6. *Perception of second subproblem.* In the course of the exploratory experiments, buckling of the upper flange appeared forcefully as a problem with no precedent in previous experience or knowledge of beams.
7. *Second act of insight.* The engineers realized that the rectangular shape was preferable to minimize the effects of sectional distortion before buckling and to facilitate thought and analysis of the buckling problem. Alternatively this step might be looked upon as a third setting of the stage preparatory to the next act of insight.
8. *Third act of insight.* Fairbairn (or Hodgkinson) saw that buckling of the upper flange can be eliminated by making the flange of longitudinal cells (the cellular principle).[194]

With the acts of insight (7) and (8), the buckling problem (6) and to some extent the lack-of-knowledge problem (4) were solved, and with them the basic problem (1) of how to bridge the Menai Straits. One step remained:

9. *Critical revision.* Fairbairn's scale-model tests and Hodgkinson's general experiments provide the detailed information needed to adapt the cellular design to the full-scale conditions of the bridges. They thus supplied the remaining knowledge required for the complete solution of problem (4) and hence of problem (1). (One could argue, however, that the buckling of the sides in Fairbairn's tests also entailed a subsequence of problem perception, stage setting, and insight.)

The picture that thus emerges is of an inward-moving sequence of problems within problems. Each problem after the first is perceived as a result of stage setting or insight aimed at solving the preceding one. When insight for solution of the innermost problem is found, the engineers can proceed outward through the sequence, solving each problem in turn and hence the original one. Considerable critical revision may be required as this outward movement proceeds.

We can attempt alternatively to analyze the events according to Usher's own conception of a strategic invention, with four major steps and subsidiary elements of novelty feeding into them. One possible arrangement is to regard item (4)—lack of knowledge—as the major problem and items (1), (2), and (3) as part of the perception of that problem. Items (5), (6), and (7) are then all part of setting the stage for insight (8), which supplies the major item of knowledge that was lacking. Critical revision then follows as in (9). This arrangement seems less satisfactory than the one proposed here in that it downplays some important steps and overemphasizes the insight into the cellular principle, which turned out in the end to be a less crucial piece of knowledge than was originally thought. We thus see Usher's four steps as supplying a useful schema for analyzing the creative events in the design of the tubular bridges, but perhaps not in quite the way that Usher himself envisioned.

The concept of problems within problems, connected by stage setting and acts of insight, arises naturally out of a consideration of the tubular-bridge episode. We suspect, however, that it may have much wider applicability.

Notes

1. For a discussion of a range of possible focusing devices, see N. Rosenberg, "The Direction of Technological Change," *Economic Development and Cultural Change* (October 1969). See also E. A. Wrigley, "The Supply of Raw Materials in the Industrial Revolution," *Economic History Review* 15, no. 1 (1962).

2. For a development of this theme in nineteenth-century America, see N. Rosenberg, "Technological Change in the Machine Tool Industry, 1840–1910," *Journal of Economic History* 23, no. 4 (1963).

3. J. H. Clapham, *An Economic History of Modern Britain* (Cambridge, 1963), 2:22.

4. Phyllis Deane and W. A. Cole, *British Economic Growth, 1688–1959* (Cambridge, 1962), p. 225. The growing importance of the iron industry in the British economy can be gauged from the fact that the gross output of the industry (less coal and imported ore) was estimated to amount to 3.4 percent of GNP in 1818 and 6.2 percent in 1851. Most of this growth in fact occurred during the 1840s. The figure reached a peak in 1871 when it amounted to 11.6 percent. Ibid., p. 226.

5. Pig iron contains 2 to 5 percent carbon alloyed with the iron and is the direct product of the smelting of iron ore in a blast furnace. It melts at a relatively low temperature (about 2,200°F) and when melted is readily cast into molds; hence its alternative name of cast

iron. It is very strong in compression but weak in tension in the ratio of about 6 to 1. Wrought iron, by contrast, contains less than about 0.1 percent of alloyed carbon (plus 1 to 2 percent of mechanically mixed impurities, called slag). By the nineteenth century it was being made from pig iron by an additional process that removed most of the carbon. It has a higher melting point (about 2,700°F). When heated to redness it becomes malleable—which cast iron does not—and can be hammered or rolled into shape. It is, however, unsuited for casting. Wrought iron is strong in tension as well as compression; in this case the ability to withstand tension is slightly the greater in the ratio of about 6 to 5. At the time in question cast iron had been available in sizable quantities for centuries; wrought iron was only then becoming obtainable in industrially significant amounts as a result of the development of the puddling process for removing the carbon from pig iron by Henry Cort in 1784 and its subsequent improvement, notably by Joseph Hall in 1839. See H. R. Schubert, "Extraction and Production of Metals: Iron and Steel," in C. Singer et al., eds., *A History of Technology* (Oxford, 1958), vol. 4. Steel, which is now the predominant structural material, has a carbon content intermediate between cast and wrought iron. It did not come into structural use until later and plays no role in our story.

6. Deane and Cole, *British Economic Growth*, pp. 231-32.

7. S. Smiles, *The Life of George Stephenson and of His Son Robert Stephenson*, rev. ed. (New York, 1868), pp. 431-32.

8. Robert Stephenson (1803-1859) was the son of the pioneer railway inventor and builder George Stephenson and was himself one of the most eminent railway engineers of his day. For the story of his life, see Smiles, *Life of Stephenson*, and L. T. C. Rolt, *The Railway Revolution: George and Robert Stephenson* (New York, 1962).

9. The two bridges, both of the tubular type, were about fifteen miles apart on the London-to-Dublin rail route. The Conway Bridge was built across the mouth of the Conway River in northern Wales. It was a low-level bridge having two lines, each with a single span of somewhat smaller dimensions than the largest of the several spans of the high-level Britannia Bridge. Its design and construction overlapped but slightly preceded that of the Britannia Bridge. Construction of the Conway Bridge began in April 1847, and the first train passed through the first tube in April 1848. The first tube of the Britannia Bridge was opened to traffic in March 1850 and the second tube in October 1850. Since the Britannia Bridge presented all the problems of the Conway Bridge and had many unique problems of

its own, we have concentrated our discussion upon the former. The building of the Conway Bridge became, as Rolt put it, "the scene of a dress rehearsal for the larger and more difficult undertaking at the Menai." Rolt, *Railway Revolution*, p. 307.

10. The most authoritative account is by Stephenson's assistant Edwin Clark, *The Britannia and Conway Tubular Bridges* (London, 1850). We have drawn heavily upon Clark's treatment for factual information. For the account by Stephenson's associate William Fairbairn, see Fairbairn, *An Account of the Construction of the Britannia and Conway Tubular Bridges* (London, 1849). Other useful accounts by contemporaries are the book by G. Drysdale Dempsey, *Tubular and Other Iron Girder Bridges* (London, 1850), and the pamphlet by Edwin Clark's brother and fellow engineer Latimer Clark, *General Description of the Britannia and Conway Tubular Bridges*, 5th ed. (London, 1850).

11. At an earlier stage, in 1839, George Stephenson had proposed that the railway make use of Telford's suspension bridge by disengaging the locomotives and using horses to haul the railway carriages across.

12. "Although some suspension bridges were still giving good service in the 1840s, many had collapsed through the buffeting effect of winds, or even through oscillations built up by the impact of marching troops or droves of cattle. [Captain Samuel] Brown had built many, and had quite a number of failures, including his bridge of 449-ft. span across the Tweed at Berwick; this he completed in 1820, but it was blown down only six months later. In 1830 he had even built one to carry the Stockton and Darlington Railway across the Tees, only to find that the train sank down and raised before and behind it a wave in the deck, which racked the bridge to pieces in a few years." S. B. Hamilton, "Building and Civil Engineering Construction," in Singer et al., *History*, 4:460. The collapse in May 1847 of the Dee Bridge (a trussed-girder bridge designed by Robert Stephenson) led to a thorough investigation by a parliamentary commission of the application of iron in railway bridges. See Parliamentary Papers, *Report of the Commission Appointed to Inquire into the Application of Iron to Railway Structures* (1849).

13. The truss bridge, which is the conventional long-span railway bridge today, was only at that time being developed, primarily in the United States and Russia; see note 112.

14. Clark, *Britannia and Conway Bridges*, p. 8.

15. Ibid., p. 29.

16. Some of the matters in this section are discussed in less detail and for different purposes by S. Timoshenko, *History of Strength of Materials* (New York, 1953), pp. 123-129, 156-62, and by I. Todhunter and K. Pearson, *A History of the Theory of Elasticity* (Cambridge, 1886), 1:785-801.

17. H. S. Smith, "Bridges and Tunnels," in Singer et al., *History*, 5:505.

18. Fairbairn, *Account*, p. 185.

19. For a general discussion of iron beams in this period, including the important consequences of the differences between cast and wrought iron, see R. A. Jewett, "Structural Antecedents of the I-Beam, 1800-1850," *Technology and Culture* 8 (1967).

20. In any so-called thin-walled structure, of which the bridge tubes are an example, the thickness of the walls is very small compared with the overall dimensions of the cross-section. As a thin-walled structure in which the walls act as primary load-resisting elements (rather than as just a simple covering), the Britannia Bridge was, in fact, an early example of the kind of structure essential to the modern-day airplane and submarine. The improvement in the knowledge of thin-walled structures as it was transferred back and forth between civil engineering, airplanes, and submarines would be a valuable topic for study.

21. Clark, *Britannia and Conway Bridges*, p. 27.

22. Ibid.

23. As it turned out, the weight of the bridge was considerably the larger—three times the weight of an equal-length train composed entirely of locomotives.

24. Timoshenko, *History*, p. 212. See also T. M. Charlton, "Contributions to the Science of Bridge-Building in the Nineteenth Century by Henry Moseley, Hon. Ll.D., F.R.S. and William Pole, D.Mus., F.R.S.," *Notes and Records of the Royal Society of London* 30 (1976). For Moseley's own writings, see, for example, H. Moseley, *The Mechanical Principles of Engineering and Architecture*, 1st American ed. (New York, 1856). Edwin Clark cites Moseley's writings at numerous places in his volumes on the construction of the bridges.

25. Clark, *Britannia and Conway Bridges*, p. 32.

26. For a full account of Fairbairn's life, see his partial autobiography *The Life of Sir William Fairbairn, Bart.*, ed. William Pole (London, 1877). See also Timoshenko, *History*, pp. 123-26.

27. For a brief account of Hodgkinson's life, see Timoshenko, *History*, pp. 126–29.

28. The general situation in England in the mid-nineteenth century is exemplified by R. H. Parsons, *A History of the Institution of Mechanical Engineers* (London, 1947), pp. 106–109: "Prior to 1853 the papers read before the Institution had been almost exclusively of a descriptive or practical kind, but in that year this custom was departed from by a mathematical disquisition on the action of centrifugal pumps by Professor Andrew J. Robertson. This marked the first appearance of the differential calculus in the Proceedings. No discussion apparently took place on the paper, possibly because it was somewhat above the heads of the ordinary members." Differential equations and their solutions are employed, however, in a number of places in the account by Clark, *Britannia and Conway Bridges*, pp. 268–74, 280-92, 739, 772–86.

29. Clark, *Britannia and Conway Bridges*, pp. 83–99, 116–26.

30. For an analytical discussion of parameter variation, see W. G. Vincenti, "A Case Study in Technological Methodology: The Air-Propeller Tests of W. F. Durand and E. P. Lesley" (in press).

31. Fairbairn was a notable contributor to boiler design and boiler-making technique. See *Life of Fairbairn*, chap. 16, p. 316.

32. Clark, *Britannia and Conway Bridges*, p. 83.

33. Ibid., p. 144.

34. Timoshenko, *History*, p. 157.

35. Clark, *Britannia and Conway Bridges*, p. 104.

36. Ibid., p. 116. Considerations were further complicated in the initial stages by an erroneous idea of Stephenson that by varying the vertical dimension of the tube suitably along its length, the material might be put entirely in tension. This condition was thought desirable because of a supposed inability of wrought iron to withstand compression. Fairbairn, *Account*, pp. 5–11. Obviously knowledge of beam action and the properties of wrought iron still left much to be desired.

37. Clark, *Britannia and Conway Bridges*, p. 116.

38. Adoption of the rectangular shape is frequently attributed simply to a greater strength of that section in comparison with the others (see, for example, Smith, "Bridges and Tunnels," p. 504). Stephenson himself made a statement to this effect in a report to the directors of the railway, perhaps to simplify matters for his non-

professional audience (note 47). Clark's explanation, however, makes it clear that other considerations were involved.

39. Bending of the sides (or webs) of the beam also gives rise to varying tension and compression forces at different vertical positions within the sides. Since the (usually) thinner sides are a relatively small part of the beam and relatively ineffective to resist bending, these forces may be neglected for present purposes and are not shown in figure 6b.

40. "On the Transverse Strain, and Strength of Materials," *Memoirs of the Literary and Philosophical Society of Manchester*, 2d series, vol. 4 (1824); "Theoretical and Experimental Researches to Ascertain the Strength and Best Forms of Iron Beams," *op. cit.*, vol. 5 (1831). An abstract of Hodgkinson's work on cast-iron beams was published in Peter Barlow, *The Strength of Timber, Cast-Iron, Malleable Iron, and Other Materials*, rev. ed. (London, 1845), pp. 226–44.

41. "Experimental Researches on the Strength of Pillars of Cast Iron, and other Materials," *Philosophical Transactions of the Royal Society of London* 130 (1840). The results of this work, as well as of Hodgkinson's earlier work on beams, were included in *Experimental Researches on the Strength and Other Properties of Cast Iron*, which Hodgkinson wrote as part 2 (published 1846) to the fourth edition of the widely used book by Thomas Tredgold, *Practical Essay on the Strength of Cast Iron* (London, 1842).

42. Timoshenko, *History*, p. 208.

43. The situation is a bit more complicated for the circular and elliptical tubes, but the basic notions and final outcome are the same.

44. This technological insight was recorded at least as early as Galileo, who wrote, "I wish to discuss the strength of hollow solids, which are employed in art—and still oftener in nature—in a thousand operations for the purpose of greatly increasing strength without adding to weight; examples of these are seen in the bones of birds and in many kinds of reeds which are light and highly resistant both to bending and breaking. For if a stem of straw which carries a head of wheat heavier than the entire stalk were made up of the same amount of material in solid form it would offer less resistance to bending and breaking. This is an experience which has been verified and confirmed in practice where it is found that a hollow lance or a tube of wood or metal is much stronger than would be a solid one of the same length and weight, one which would necessarily be thinner; men have discovered, therefore, that in order to make lances strong

as well as light they must make them hollow." *Dialogues Concerning Two New Sciences*, trans. H. Crew and A. de Salvio (New York, 1933), p. 150.

45. Clark, *Britannia and Conway Bridges*, p. 145.

46. According to a later statement by Fairbairn, *Account*, p. 21, "It is from this period that I date the disappearance of almost every difficulty respecting the construction and ultimate formation of the Britannia and Conway tubes."

47. The reports are included in Clark, *Britannia and Conway Bridges*, pp. 135–54. Stephenson's views of the experiments are of interest: "In the course of the experiments, it is true, some unexpected and anomalous results presented themselves; but none of them tended, in my mind, to show that the tubular form was not the very best for obtaining a rigid roadway for a railroad over a span of 450 feet, which is the absolute requirement for a bridge over the Menai Straits.

"The first series of experiments was made with plain circular tubes, the second with elliptical, and the third with rectangular. In the whole of these this remarkable and unexpected fact was brought to light, viz. that in such tubes the power of wrought-iron to resist compression was much less than its power to resist tension, being exactly the reverse of that which holds with cast-iron: for example, in cast-iron beams for sustaining weight, the proper form is to dispose of the greater portion of the material at the bottom side of the beam, whereas with wrought-iron, these experiments demonstrate, beyond any doubt, that the greater portion of the material should be distributed on the upper side of the beam. [Stephenson appears here to regard resistance to compression as depending only on the material itself and to ignore the structural-type of failure caused by buckling. This seeming misunderstanding is curious since Fairbairn and Hodgkinson in their reports clearly understood that buckling had to be eliminated in order to utilize the full strength of the material. Stephenson probably still harbored the misconception that wrought iron was inherently weak in compression. See notes 36 and 169.] We have arrived therefore, at a fact having a most important bearing upon the construction of the tube; viz. that rigidity and strength are best obtained by throwing the greatest thickness of material into the upper side. [Later results showed that Stephenson here overemphasized the amount of additional material needed on the upper side when wrought iron is used.]

"Another instructive lesson which the experiments have disclosed is, that the rectangular tube is by far the strongest, and that the circular and elliptical should be discarded altogether. [See note 38.]

"This result is extremely fortunate, as it greatly facilitates the mechanical arrangements for not merely the construction, but the permanent maintenance of the bridge.

"We may now, therefore, consider that two essential points have been finally determined,—the form of the tube, and the distribution of the material.

"The only important question remaining to be determined, is the absolute ultimate strength of a tube of any given dimensions. This is, of course, approximately solved by the experiments already completed."

Stephenson's report was given wide circulation by being reprinted in *The Artizan* (March 1846):160.

48. Clark, *Britannia and Conway Bridges*, p. 34.

49. This view and its aftermath are an excellent example of D. A. Schon's illuminating idea of displacement of concepts whereby new concepts are formed through the "displacement of old concepts to new situations resulting in extension of the old." *Invention and the Evolution of Ideas* (London, 1967), p. 34; see also pp. x, 53. In this case the old concept of the buckling of columns was applied to the new situation of the compression flange of a thin-walled beam. This displacement extended the old concept, leading eventually to the new and broader concept of the buckling (or elastic instability) of structural elements generally.

50. This is the ratio given by Clark, *Britannia and Conway Bridges*, p. 163, following a detailed listing of the areas of the various components. See also Fairbairn, *Account*, p. 253. Timoshenko, *History*, p. 158, gives the smaller value 5:3 (equivalent to 1.67:1). This is actually the value appropriate to the cellular-flanged tube of figure 7 rather than to the one-sixth-scale model; see Fairbairn, *Account*, pp. 20, 22, 70.

51. Clark, *Britannia and Conway Bridges*, p. 184.

52. Fairbairn, *Account*, p. 145.

53. Or webs in the case of the tubular beam.

54. Timoshenko, *History*, p. 143.

55. Clark, *Britannia and Conway Bridges*, p. 134.

56. Ibid., p. 148.

57. Ibid., pp. 318–64.

58. Ibid., pp. 397–419.

59. Timoshenko, *History*, p. 162.

60. Ibid., p. 299.

61. Clark, *Britannia and Conway Bridges*, sec. IB; Fairbairn, *Account*, pp. 284-88.

62. According to Todhunter and Pearson, *History of Elasticity*, p. 791, these experiments on rivets "seem to be among the earliest of the kind."

63. The statement is sometimes made (by Timoshenko, *History*, p. 159, for example) that the final design was obtained by scaling up the model results on the assumption that for geometrically similar tubes the carrying capacity varies as the square of the length of the tube (verified by Hodgkinson's tests) and the weight varies as the cube. These assumptions were in fact used as a basis for estimating the failing loads for the final designs (see Clark, *Britannia and Conway Bridges*, pp. 743-49), but the design process itself was not that simple.

64. Clark, *Britannia and Conway Bridges*, pp. 725-42.

65. Ibid., pp.743-87. Clark indicates in his preface (p. vii) that the analytical portions of his account (sections III and VIII) were contributed by Pole. For particulars on Pole and his work, see Charlton, *Contributions*, pp. 174-76.

66. Comparative figures are as follows (the first figure in each case is for the Britannia tube and the second for the Conway): longest span—460 feet, 400 feet; greatest height of tubes—30 feet, 25.5 feet; width of tubes—14 feet, 8 inches, 14 feet, 8 inches; Weight of single largest tube—1,400 tons, 1,146 tons; total weight of iron structure—10,570 tons, 2,892 tons. Fairbairn, *Account*, pp. 184-85.

67. Clark, *Britannia and Conway Bridges*, pp. 511-12.

68. Ibid., pp. 515-17.

69. Timoshenko, *History*, p. 161.

70. Ibid., p. 143.

71. Clark, *Britannia and Conway Bridges*, pp. 288-90.

72. Ibid., pp. 464-66.

73. Ibid., pp. 766-67.

74. Ibid., pp. 490-94, 704-05, 762-85. See also Timoshenko, *History*, pp. 160-61, for a brief explanation.

75. Todhunter and Pearson, *History of Elasticity*, pp. 787–88, say that portions of the theoretical calculations "possibly appear here for the first time." The matters cited, however, are details of secondary significance.

76. Clark, *Britannia and Conway Bridges*, p. 762.

77. The weight of the tubes constituted at least three-fourths of the total load; see note 23.

78. R. Stephenson, "Iron Bridges," *Encyclopaedia Britannica*, 8th ed. (Boston, 1856), 12:608; see also Clark, *Britannia and Conway Bridges*, p. 552.

79. Clark, *Britannia and Conway Bridges*, pp. 557, 579.

80. Ibid., p. 512.

81. Ibid., p. 560; see also p. 511.

82. Ibid., pp. 470, 718.

83. Ibid., pp. 469–70.

84. Ibid., sec. 5, chaps. 1, 2.

85. The copy of Fairbairn's account of the bridges in the Timoshenko Collection at Stanford University was purchased from a bookseller in Manchester and appears to have been Hodgkinson's own copy. It has penciled marginal notes, some pointedly critical, that could have been written only by Hodgkinson.

86. Timoshenko states in this regard, *History*, p. 193, "Fairbairn's experiments first drew the attention of engineers to the importance of the stability question in designing compressed iron plates and shells."

87. Fairbairn, *Account*, p. 185.

88. The fastening together of metal pieces by rivets is now a less familiar process than it once was, having been superseded in steel structures largely by welding and bolting. It involves punching a hole through overlapping pieces of metal, placing a red-hot rivet (a pin consisting of a cylindrical shank and a head) into the hole, and forming a second head on the opposite side of the plates from the first, thus holding the plates together. The last step is referred to variously as "driving," "closing," "heading," or "riveting." The word "riveting" is also used to refer to the entire procedure.

89. Clark, *Britannia and Conway Bridges*, p. 581.

90. See H. W. Dickinson, "Richard Roberts, His Life and Inventions," *Transactions of the Newcomen Society* 25 (1945–1947).

91. L. T. C. Rolt, *A Short History of Machine Tools* (Cambridge, Mass., 1965), pp. 107–08. See also Clark, *Britannia and Conway Bridges*, p. 650.

92. Fairbairn, *Account*, p. 154.

93. Clark, *Britannia and Conway Bridges*, p. 632.

94. Ibid., pp. 628-32; Fairbairn, *Account*, p. 154.

95. Clark, *Britannia and Conway Bridges*, p. 666.

96. Ibid., p. 626.

97. The masonry bridge towers were completed from designs that had assumed chains, which explains their seemingly excessive height above the tubes (see figure 1).

98. The tubes of the shorter side spans, which were almost entirely over land, were fabricated in place on stagings built up from the ground.

99. Ibid., p. 614.

100. For a full account of this dramatic incident, see Rolt, *Railway Revolution*, p. 314.

101. Todhunter and Pearson, *History of Elasticity*, p. 795.

102. *Report of the Commission Appointed to Inquire into the Application of Iron to Railway Structures* (1849), pp. 365–66. Hodgkinson was a member of the commission from 1847 to 1849.

103. Clark, *Britannia and Conway Bridges*, p. 716.

104. Ibid., p. 530.

105. W. S. Jevons, *The Coal Question* (London, 1865), p. 97.

106. Timoshenko, *History*, chap. 6.

107. Ibid., p. 162.

108. D. J. de S. Price, "Is Technology Historically Independent of Science? A Study in Statistical Historiography," *Technology and Culture* 6 (1965).

109. Smiles, *Life of Stephenson*, p. 475, and Stephenson, "Iron Bridges," pp. 606–10. In the Egyptian bridges the vertical dimension of the tubes was less than that of a train, and the railway was laid on top of the tubes rather than through the interior.

110. Smiles, *Life of Stephenson*, pp. 475–84. See also J. Hodges *Construction of the Great Victoria Bridge* (London, 1860).

111. Stephenson, "Iron Bridges," p. 609, and W. J. M. Rankine, *Manual of Civil Engineering*, 8th ed., rev. (London, 1872), pp. 531–34.

112. In the rapid development of iron railway bridges in the 1840s, American and Russian engineers appear to have thought primarily in terms of trusses and British engineers in terms of riveted-plate structures. This difference raises interesting questions bearing on the origins and diffusion of technical innovations and needs more detailed study than it has so far received. Both C. Condit, "Buildings and Construction," in M. Kranzberg and C. W. Pursell, Jr., eds., *Technology in Western Civilization* (New York, 1967), 1:373, 383, and Timoshenko, *History*, pp. 183–84, attribute the American and Russian concentration on iron trusses to an extensive earlier experience with timber trusses, which stemmed in turn from an abundance of wood and from other economic factors peculiar to the United States and Russia. The evidence of the present paper strongly suggests that the British emphasis on plate structures was conditioned by a growing availability of wrought iron in Britain and by the associated prior experience with riveted-plate ship construction. With a few minor exceptions British engineers began to use iron trusses only in the 1850s following development of the Warren truss between 1846 and 1848. Condit, "Buildings," p. 386; Timoshenko, *History*, p. 186. See also T. M. Charlton, "Theoretical Work," in A. Pugsley, ed., *The Works of Isambard Kingdom Brunel* (Bristol, 1976), pp. 196–8 and P. S. A. Berridge, *The Girder Bridge* (London, 1969), chap. 25. The first publications on the analysis of trusses appeared with S. Whipple's book, *An Essay on Bridge Building*, in the United States in 1847 and a journal article by D. J. Jourawski in Russia in 1850. Timoshenko, *History*, pp. 185, 189. As with many new alternative technologies, however, it was not immediately obvious which type of bridge was preferable on either technical or economic grounds. Even if the early publications on trusses had been available to Stephenson and Fairbairn in 1845, they might understandably have opted for the tubular beam. Experience and technical information weigh heavily with engineers in making major design decisions, especially where technical failure may lead to considerable loss of life.

113. Clark, *Britannia and Conway Bridges*, p. 524.

114. Patent No. 11,401 dated October 8, 1846. For the text of this patent, see W. Newton, *London Journal of Arts, Sciences and Manufactures* (London, 1847), 31:9-12.

115. *Life of Fairbairn*, p. 213.

116. R. J. M. Sutherland, "The Introduction of Structural Wrought Iron," *Transactions of the Newcomen Society* 36 (1963-1964):80. See also *Report of the Commissioners Appointed to Inquire into the Application of Iron to Railway Structures* (1849), pp. 291–96, appendix 7, plates 7, 7*.

117. See note 151.

118. See also Jewett, "Structural Antecedents," p. 361.

119. Sutherland, "Introduction," p. 82. See also Fairbairn's *On the Application of Cast and Wrought Iron to Building Purposes*, 2d ed. (London, 1857-1858), pp. 81-88, 255-58.

120. *Life of Fairbairn*, p. 213.

121. Brunel in the early 1840s "carried out large-scale experiments on the strength of plate girders [of single-web section] and used novel forms of compression flange including the tube and variants thereon." Charlton, "Theoretical Work," p. 191. See also L. T. C. Rolt, *Isambard Kingdom Brunel* (London, 1957), p. 180. These experiments are discussed at some length in Clark's *Britannia and Conway Bridges*, pp. 437–41.

122. For Fairbairn's more definite but not disinterested statement on this, see *Application of Cast and Wrought Iron*, p. 259.

123. For discussion of the Chepstow, Saltash, and Forth bridges, see Berridge, *Girder Bridge*, chaps. 9, 10, 12.

124. *The Mechanics' Magazine*, n.s. 4 (September 21, 1860):181. A. W. Skempton suggests that the knowledge from the tubular bridges may have had application also in the design of the wrought-iron girders of the historically important Boat Store in H.M. Dockyard, Sheerness. "The Boat Store, Sheerness (1858–1860) and Its Place in Structural History," *Transactions of the Newcomen Society* 32 (1956-1960):73.

125. S. Giedion, *Space, Time and Architecture*, 5th ed. (Cambridge, Mass., 1973), p. 194.

126. Condit, "Buildings," p. 376.

127. *Life of Fairbairn*, p. 321.

128. Fairbairn, *Useful Information for Engineers, Second Series*, 2d ed. (London, 1867), p. 139.

129. Ibid. Chain riveting, which is used to join plates subject to tension, describes a chainlike arrangement of rivets one after the other along a line in the direction of the tensile force.

130. Ibid., p. 147. See also the opening sentence of the chapter: "These structures are identical in principle with the tubular bridges over the Conway and Menai Straits, and present additional examples of the advantages which may yet be derived from a judicious combination of wrought iron plates in constructions requiring security, rigidity, and great strength." Ibid., p. 138. See also Fairbairn, "Tubular Wrought Iron Cranes," *Proceedings of the Institution of Mechanical Engineers* (1857), p. 93.

131. Rolt, *Short History of Machine Tools*, p. 120. In describing Whitworth's machine tools in 1862, D.K. Clark made the following observations: "Trying back to first principles, Messrs. Whitworth & Co. introduced the method of the hollow, cored, or box casting for the frames of machine-tools, on a principle which may at once be illustrated and enforced by the familiar example of the goose-quill, in which the maximum of strength and stiffness is combined with the minimum consumption of material: The material is removed from the centre and distributed circumferentially, as in a tube or a box, round or square. The capacity of the box casting for resisting strains of every variety—longitudinal, transverse, or torsional—is amazing; and, if it is not the most useful thing those exhibitors have done, it is certainly one of the happiest. The same constructors, in pursuance of the principle of simplicity and directness, have eschewed fragmentary or pieced framework, and, as far as practicable, have cast the lesser upon the larger members. In this matter, as in the construction of hollow framing, other exhibitors have adopted the same practice." D. K. Clark, *The Exhibited Machinery of 1862* (London, 1862), p. 129.

132. Fairbairn described his experience as an iron shipbuilder: "I was the first to commence iron shipbuilding in London, and, I believe, was the second to send an iron vessel to sea. From 1829-30 to 1848 I built upwards of 120 iron vessels, some of them upwards of 2000 tons burden, and nine of which were built in sections at Manchester, and the remainder on the banks of the Thames at Millwall." *Useful Information, Second Series*, p. 79. For a detailed account, see *Life of Fairbairn*, especially chaps. 9, 10, 19.

133. See E. C. Smith. *A Short History of Naval and Marine Engineering* (Cambridge, 1938), esp. chap. 7.

134. Fairbairn, *Useful Information, Second Series*, p. 101. Later Fairbairn amplified: "If we take a vessel such as we have described . . . , we shall approximate nearly to the facts by treating it as a simple beam; actually a vessel is placed in this position, either when supported at each end by two waves or when rising on the crest of another wave, supported at the centre, with the stem and stern partially suspended. Now, in these positions the ship undergoes alternately a strain of compression and a strain of tension along the whole section of the deck, corresponding with equal strains of tension and compression along the whole section of the keel, the strains being reversed according as the vessel is supported at the ends or the centre. These are, in fact, the alternate strains to which every long vessel is exposed, particularly in seas where the distance between the crests of the waves does not exceed the length of the ship." Ibid., pp. 103-04. See also Fairbairn, "The Strength of Iron Ships," *Transactions of the Institution of Naval Architects* (1860):72.

135. Fairbairn, *Useful Information for Engineers*, 4th ed. (London, 1864), p. 239.

136. Fairbairn, *Useful Information, Second Series*, pp. 100-37. "The surveyors of Lloyd's, most excellent, well-meaning, gentlemenly men as they are, may say what they please, but I have no hesitation in stating that their regulations are very defective and require immediate revision." Ibid., p. 125.

137. "If I am correct in treating iron vessels in the light of simple girders, I shall be able to show a better disposition of material, calculated to remedy present defects, and greatly increase the strength of vessels, without any great increase of cost, to resist transverse strains. If I were proceeding upon theoretical considerations, the results I have stated might be doubted; but we have a sufficient number of experiments upon hollow wrought-iron girders to calculate the strength and resisting powers of ships to transverse strain, with a near approximation to accuracy in the results." Ibid., p. 111.

138. Fairbairn, *Treatise on Iron Ship Building: Its History and Progress* (London, 1865), p. 214.

139. "The result of this application, with the longitudinal bulkheads, constitutes the enormous strength of this magnificent vessel, proving the importance of the cellular system for vessels of large tonnage. It combines lightness with strength, and double sheathing gives immense rigidity to the construction." *Useful Information, Second Series*, pp. 119-20. For further treatment of cellular design, see ibid.,

pp. 125–37. A later observer states: "Mr. Brunel was not content . . . to limit the application of the cellular system to the lower part of the ship, but extended it to the upper deck, which is nothing more nor less than an imitation of the top flange of the Menai Bridge." *Great Industries of Great Britain* (London, 1877–1880?), vol. I, p. 266.

140. A. M. Robb, "Ship-Building," in Singer et al., *History* 5:362–63. While the ship was under construction *The Artizan* observed that "the cellular construction of the sides, the bottom, and the main deck, converts the whole into one immense wrought-iron tubular bridge or beam." *The Artizan* (July 1856):145. In a later article it was suggested that the ship's first project after launching should be the laying of a transatlantic cable—a project that the ship actually undertook, years later, after its huge coal consumption (among other difficulties) had rendered it a financial failure. *The Artizan* (October 1856):217–18.

141. Robb, "Ship-Building," p. 364.

142. Ibid. Fairbairn's concern with riveting techniques went back to his earlier preoccupation with boilermaking. See *Life of Fairbairn*, pp. 163–64.

143. "An Experimental Inquiry into the Strength of Wrought-Iron Plates and their Riveted Joints as applied to Ship-building and Vessels exposed to severe strains," *Philosophical Transactions of the Royal Society of London* 140 (1850); Todhunter and Pearson, *History of Elasticity*, pp. 800–01, question various of Fairbairn's results on riveted joints but conclude that he "can however lay claim to being the earliest and for many years the only experimenter in this field."

144. Fairbairn, *Useful Information, Second Series*, pp. 117, 122.

145. See also Fairbairn, *Account*, p. 284, and *Treatise on Iron Ship Building*, p. 40. Clark, *Britannia and Conway Bridges*, pp. 518–19, did not agree with Fairbairn about the effectiveness of chain riveting, maintaining that experiments showed a zig-zag arrangement to be stronger (cf. note 129). Lloyd's and other underwriters' rules, however, subsequently recommended chain riveting; see B. D. Stoney, *The Strength and Proportions of Riveted Joints* (London, 1885), pp. 15–19.

146. *Life of Fairbairn*, p. 342.

147. Rolt, *Brunel*, chap. 14.

148. *Report of the Commissioners.*

149. *Supplement to the Theory, Practice, and Architecture of Bridges*, ed. G. R. Burnell (London, 1850), pp. 61–92.

150. See his several series and editions (some previously cited) of *Useful Information for Engineers* (London, 1856 et seq.), *Second Series* (London, 1860 et seq.), *Third Series* (London, 1866). See also *Application of Cast and Wrought Iron*, and *Treatise on Iron Ship Building*.

151. See, for example, Fairbairn's "On Tubular Girder Bridges," *Minutes of the Proceedings of the Institution of Civil Engineers*, Session 1849–1850 (London, 1850), vol. 9. This paper dealt primarily with the safety of the controversial Torskey Bridge, a hollow-girder bridge built by John Fowler over the River Trent. The explicit basis for the assessment, however, was the knowledge gained on the Britannia and Conway bridges. Following the presentation there was lengthy discussion and dispute, lasting several evenings and involving numerous engineers. Matters of discussion included the validity of Fairbairn's formula for calculating the strength of hollow girders (derived from the tubular-bridge experience), the omission of the sides of the girders from the strength calculations, the influence of the continuity of the girders across the center support, and the effect of impact and vibration from moving loads. Reference to the Britannia and Conway bridges appears throughout the discussion.

152. *Minutes*, Session 1850–1851 (London, 1851), 10:175, 182, 184.

153. T. Tate, *Strength of Materials* (London, 1850) (inspired almost entirely by the bridge experience); B. Baker, *On the Strengths of Beams, Columns, and Arches* (London, 1870), pp. 102, 187–88; J. Anderson, *The Strength of Materials and Structures* (London, 1872), pp. 159–60, 202–03.

154. For example, W. J. M. Rankine, *Manual of Applied Mechanics*, 6th ed., rev. (London, 1872), pp. 364–68, and *Manual of Civil Engineering*, 8th ed., rev. (London, 1872), pp. 237–38, 522–23, 527–38.

155. E. H. Salmon, *Columns* (London, 1921), pp. 177–79, 182.

156. Timoshenko, *History*, pp. 123, 160, 191–93.

157. Todhunter and Pearson, *History of Elasticity*, p. 795.

158. C. Tomlinson, ed. (London, 1868), 1:239–52.

159. R. Hunt, ed., 6th edition (London, 1867), 3:949–57. The article in the fourth edition (Boston, 1853), 1:668–87, the last published under the authorship of Dr. Andrew Ure, also contains a

lengthy polemic giving Fairbairn almost sole credit for the successful construction of the Britannia and Conway bridges. The article in this edition is, in fact, entitled "Fairbairn's Tubular Bridges."

160. Stephenson, "Iron Bridges," pp. 575-610.

161. E. H. Knight, (New York, 1877), 3:2646-47.

162. For a discussion of this complex matter see Timoshenko, *History*, pp. 146, 160-61.

163. For example, L. T. C. Rolt, *Victorian Engineering* (London, 1970), p. 30, states that the work of the bridge engineers "not only achieved their immediate purpose but laid the foundations of modern structural engineering theory." Even if one interprets "foundations" very loosely, this is hardly correct. The foundations of structural theory were laid, primarily in France, well before the tubular bridges were conceived; see Timoshenko, *History*, chaps. 3-5.

164. Ibid., p. 123. Todhunter and Pearson, *History of Elasticity*, p. 795, say that "these contributions [from the bridge work] have considerable bearing on theory." Here they clearly have in mind implications for theoretical development or a quantitative basis for verifying those developments. These are not the same as contributing to the theory itself.

165. Sutherland, "Introduction."

166. S. B. Hamilton ("Building Materials and Techniques," in Singer et al., *History*, 5:472) concerns himself mainly with buildings and is hence more conservative. He writes, "In the 1850s and 1860s, while the fashion for cast-iron construction was still at its height, there occurred a considerable extension of the use of wrought iron."

167. Sutherland, "Introduction," p. 69. For a contemporary (1845) observation of the lack of experiments on the bending of wrought iron prior to this period, see Barlow, *Strength of Timber*, p. 306.

168. In ship hulls, in particular, wrought iron was used for framing, as well as plating, and this use played a significant role in the development of the wrought-iron beam. Jewett, "Structural Antecedents," pp. 350-55.

169. Sutherland, "Introduction," p. 81. Fairbairn's own words are also of interest: "Before that date, (1845-6), our knowledge of the properties of wrought iron, and its application to the useful arts, was very imperfect. It had been used in the construction of boilers, steam engines, and water wheels, from a comparatively early period,

and even at that time and for some years previous, it was making rapid progress in its application to ship-building. Its properties, distribution, and appliance to beams and bridges, were, however, unknown and unappreciated until the experiments referred to proved its superiority over every other material then known for the attainment of objects for which it has since been so largely and so extensively in demand. As a material for the construction of bridges, wrought iron was universally condemned, and some of our ablest mathematicians went so far as to prove its inefficiency in the shape of rectangular tubes composed of riveted plates, as being perfectly utopian, and, to employ the expression then made use of, "*It would crumple up like a piece of leather!*" *Application of Cast and Wrought Iron* p. 258.

170. Sutherland, "Introduction," p. 68.

171. Clark, *Britannia and Conway Bridges*, p. 693.

172. Ibid., pp. 693–94.

173. Dempsey, *Tubular and Other Iron Girder Bridges*, p. 124. See also *Civil Engineer and Architect's Journal* 12 (October 1849): 305–06.

174. Anderson's description and figure are not precisely correct. As previously mentioned (on the evidence of plate 27 of the volume of plates accompanying Clark, *Britannia and Conway Bridges*), the bottom of the original cylinder was slightly dished rather than flat. The bottom of the replacement cylinder was not a hemisphere but half a prolate ellipsoid–that is, more elongated vertically than a hemisphere. The same imprecision has been repeated by other writers.

175. Anderson, *Strength of Materials*, pp. 280–82. Anderson's treatment of the issue of crystallization as it is influenced by the shape of the casting is repeated in his Cantor lectures. See *Journal of Society of Arts* 17 (August 1869):746. Anderson was an important and influential figure in British technological affairs around the middle of the nineteenth century. For further information concerning his work, see N. Rosenberg, ed., *The American System of Manufactures* (Edinburgh, 1969), pp. 80–86 and passim.

176. *Great Exhibition 1851. Official, Descriptive, and Illustrated Catalogue. Part II. Machinery* (London, 1851), pp. 226–27.

177. A particularly interesting example is the column-of-water engine from the 1700s, a reciprocating engine much like the early steam engines but using water under pressure rather than steam. Al-

though this engine has now been almost forgotten, the analogies that it provided, as Cardwell has clearly demonstrated, probably contributed significantly to the early development of the science of thermodynamics in the work of Sadi Carnot. D. S. L. Cardwell, "Power Technologies and the Advance of Science, 1700-1825," *Technology and Culture* 6 (1965).

178. The original bridge was distorted and weakened by a fire on May 23, 1970, and had to be replaced. The fire was started by two youths who trespassed on the bridge in search of birds' nests and, to aid their search, lit torches of rolled-up newspaper. A timber roof over the tubes, which had been added for protection early in the twentieth century and which was heavily coated with tar, caught fire from one of the torches. The fire quickly spread along the entire length of the bridge, igniting a tar deposit on top of the tubes themselves. For a description of the dismantling of the damaged tubular bridge and the design and construction of the new arch bridge, see H. C. Husband and R. W. Husband, "Reconstruction of the Britannia Bridge," *Proceedings of the Institution of Civil Engineers, Part 1* (London, 1975), vol. 58.

179. For details, see Clark, *Britannia and Conway Bridges*, pp. 30-31.

180. Ibid., p. 21.

181. Smith, "Bridges and Tunnels," pp. 503-04.

182. Clark, *Britannia and Conway Bridges*, p. 50.

183. The interested reader may wish, at this point, to compare our treatment with Albert Hirschman's highly provocative discussion of the role of the "hiding hand" in *Development Projects Observed* (Washington, D.C., 1967). Hirschman argues that people characteristically underestimate their capacity for creative action. Given this systematic bias, we might regularly fail to undertake worthwhile projects were it not for the beneficent effects of the hiding hand that conceals from our view difficulties that we are capable of surmounting. In contrast with such a situation, the main figures in the Britannia Bridge episode were well aware of serious difficulties at the outset but were sufficiently confident of their ability to surmount them. At the same time the hiding hand also functioned with respect to certain unanticipated difficulties (such as the buckling problem) that the engineers proved capable of solving.

184. For extended discussion of this point, see Vincenti, "Case Study."

185. See the useful introduction by A. E. Musson to the 1970 reprint of *Life of Fairbairn* (David & Charles Reprints, Newton Abbot, 1970).

186. Waterwheels were far from an anachronism in mid-nineteenth-century Britain, fully three-quarters of a century after some of Watt's most seminal innovations. The reason we are likely to regard them as such is because of the habit, inculcated by writers of history textbooks, of drastically exaggerating the speed with which even major inventions displace the older technology. For further discussion, see N. Rosenberg, "Factors Affecting the Diffusion of Technology," *Explorations in Economic History* (Fall 1972):3–33.

187. Although Fairbairn had conducted much experimentation on the properties of cast iron early in his career, he later became increasingly concerned with the properties and the applications of wrought iron. See the appendix to Fairbairn's *Life*, which presents a chronological listing of his writings, including "An Experimental Inquiry into the Strength of Wrought-Iron Plates and their Riveted Joints as applied to Ship-building and Vessels exposed to severe strains," *Philosophical Transactions of the Royal Society of London* (1850); "On the Security and Limit of Strength of Tubular Bridges Constructed of Wrought Iron," *Memoirs of the Literary and Philosophical Society of Manchester* vol. 9 (1850); "On the Mechanical Properties of Metals as Derived from Repeated Meltings, Exhibiting the Maximum Point of Strength and the Causes of Deterioration," *Reports of the British Association* (1853); *On the Application of Cast and Wrought Iron to Building Purposes* (London, 1856); "On the Tensile Strength of Wrought Iron at Various Temperatures," *British Association Reports* 25 (1856); "On Tubular Wrought Iron Cranes," *Proceedings of the Institution of Mechanical Engineers* (1857); "The Strength of Iron Ships," *Transactions of the Institution of Naval Architects* 1 (1860); "Experiments to Determine the Effect of Vibratory Action and Long-Continued Changes of Load upon Wrought-Iron Girders," *British Association Reports* (1860); "On the Resistance of Iron Plates to Statical Pressure, and the Force of Impact by Projectiles at High Velocities," *British Association Reports* (1861); *Treatise on Iron Ship Building* (London, 1865); "On the Durability and Preservation of Iron Ships, and on Riveted Joints," *Proceedings of the Royal Society* 21 (1873). For a discussion of part of Fairbairn's work, see A.I. Smith, "William Fairbairn and the Mechanical Properties of Materials–The Effect of Repeated Loading on Strength," *The Engineer* 217 (1964):1133–36.

188. A. P. Usher, *A History of Mechanical Inventions*, rev. ed. (Cambridge, Mass., 1954), chap. 4; for an elaboration and extension of

some of the ideas, see also his "Technical Change and Capital Formation," *Capital Formation and Economic Growth* (Princeton, 1955), reprinted in N. Rosenberg, ed., *The Economics of Technological Change* (Harmondsworth, 1971).

189. "Technical Change and Capital Formation," reprinted in Rosenberg, *Economics*, pp. 43-44.

190. V. Ruttan, "Usher and Schumpeter on Invention, Innovation and Technological Change," *Quarterly Journal of Economics* 73 (1959), reprinted also in Rosenberg, *Economics.*

191. Not discussed in the main text; see Stephenson's account in Clark, *Britannia and Conway Bridges*, pp. 22-27.

192. This insight is another example of Schon's displacement of concepts; see note 49.

193. F. M. Scherer, "Invention and Innovation in the Watt-Boulton Steam-Engine Venture," *Technology and Culture* 6 (1965).

194. There was some argument between Fairbairn and Hodgkinson about which of them was responsible for this insight; see Fairbairn, *Account*, pp. 95-96, and *Report of the Commissioners Appointed to Inquire into the Application of Iron to Railway Structures*, p. 115.

Index